W9-BLJ-946

Praise for
THE CARTOON INTRODUCTION TO CLIMATE CHANGE

"Fresh! Cheeky! Accurate and inspiring! An accessible, friendly, and fun explanation of climate change—free of politics, free of jargon, and fresh with insights. Cartoons you can believe in!"

—**Jane Lubchenco**, Wayne and Gladys Valley Professor of Marine Biology, Oregon State University, and Former Administrator of the National Oceanic and Atmospheric Administration (NOAA)

"Grady Klein and Yoram Bauman are a national treasure. The economics of climate policy has never been more accessible."

—**Kevin Hassett**, Senior Fellow and Director of Economic Policy Studies, American Enterprise Institute

"*THE CARTOON INTRODUCTION TO CLIMATE CHANGE* will tickle your fancy while expanding your mind. Highly recommended."

—**Martin Weitzman**, Professor of Economics, Harvard University

"Rarely do you read books that attempt to deal with the world's biggest problems and present the information in a way that the average public can absorb it. Bravo to Yoram Bauman and Grady Klein, and thank you on behalf of everyone who is deeply concerned about this issue."

—**Mark Reynolds**, Executive Director, Citizens Climate Lobby

"Illustrated with deceptively simple black-and-white art that masterfully supports the text, this book provides a skillful tour of the issues that face our developing world and it serves as a model of how educational works of this sort should be crafted."

—***Publishers Weekly***

ABOUT ISLAND PRESS

SINCE 1984, THE NONPROFIT ORGANIZATION ISLAND PRESS HAS BEEN STIMULATING, SHAPING, AND COMMUNICATING IDEAS THAT ARE ESSENTIAL FOR SOLVING ENVIRONMENTAL PROBLEMS WORLDWIDE. WITH MORE THAN 800 TITLES IN PRINT AND SOME 40 NEW RELEASES EACH YEAR, WE ARE THE NATION'S LEADING PUBLISHER ON ENVIRONMENTAL ISSUES. WE IDENTIFY INNOVATIVE THINKERS AND EMERGING TRENDS IN THE ENVIRONMENTAL FIELD. WE WORK WITH WORLD-RENOWNED EXPERTS AND AUTHORS TO DEVELOP CROSS-DISCIPLINARY SOLUTIONS TO ENVIRONMENTAL CHALLENGES.

ISLAND PRESS DESIGNS AND EXECUTES EDUCATIONAL CAMPAIGNS IN CONJUNCTION WITH OUR AUTHORS TO COMMUNICATE THEIR CRITICAL MESSAGES IN PRINT, IN PERSON, AND ONLINE USING THE LATEST TECHNOLOGIES, INNOVATIVE PROGRAMS, AND THE MEDIA. OUR GOAL IS TO REACH TARGETED AUDIENCES—SCIENTISTS, POLICYMAKERS, ENVIRONMENTAL ADVOCATES, URBAN PLANNERS, THE MEDIA, AND CONCERNED CITIZENS—WITH INFORMATION THAT CAN BE USED TO CREATE THE FRAMEWORK FOR LONG-TERM ECOLOGICAL HEALTH AND HUMAN WELL-BEING.

ISLAND PRESS GRATEFULLY ACKNOWLEDGES MAJOR SUPPORT OF OUR WORK BY THE AGUA FUND, THE ANDREW W. MELLON FOUNDATION, BETSY & JESSE FINK FOUNDATION, THE BOBOLINK FOUNDATION, THE CURTIS AND EDITH MUNSON FOUNDATION, FORREST C. AND FRANCES H. LATTNER FOUNDATION, G.O. FORWARD FUND OF THE SAINT PAUL FOUNDATION, GORDON AND BETTY MOORE FOUNDATION, THE KRESGE FOUNDATION, THE MARGARET A. CARGILL FOUNDATION, NEW MEXICO WATER INITIATIVE, A PROJECT OF HANUMAN FOUNDATION, THE OVERBROOK FOUNDATION, THE S.D. BECHTEL, JR. FOUNDATION, THE SUMMIT CHARITABLE FOUNDATION, INC., V. KANN RASMUSSEN FOUNDATION, THE WALLACE ALEXANDER GERBODE FOUNDATION, AND OTHER GENEROUS SUPPORTERS.

THE OPINIONS EXPRESSED IN THIS BOOK ARE THOSE OF THE AUTHOR(S) AND DO NOT NECESSARILY REFLECT THE VIEWS OF OUR SUPPORTERS.

THE CARTOON INTRODUCTION TO CLIMATE CHANGE

To Jim,

Good luck

To us all...
And thanks to E3
for its
good
work!

2/24/16

EXTRA SUPER DUPER GENEROUS

WE'RE GRATEFUL TO THE FOLLOWING PEOPLE AND ORGANIZATIONS FOR HELPING TO MAKE THIS BOOK POSSIBLE, INCLUDING VIA GENEROUS CONTRIBUTIONS ON **KICKSTARTER**.

JAMES ADCOCK
BRIAN ARBOGAST
MARCIA BAKER
REBECCA TUTTLE BALDWIN
(SEE CAMEO OF MARK, PAGE 194)
THE BAUMAN FAMILY
MARY BENNETT
JARED BIBLER
NICHOLAS A. BOND
CHRISTOPHE BONTEMPS
(SEE PAGE 142)
ALEX BOZMOSKI
GLENN BRANCH
DERIK BROEKHOFF
SUSAN BROOKS
LINDA R. BROWN
RACHEL BROWN
TRUMAN BUFFETT
JEFF CASWELL
ANDREW CHRISTOPHE
MATT CLEMENTS
ANA UNRUH COHEN
(SEE PAGE 195)
KENN COMPTON
JAMES CONKLIN
EUGENE CORDERO
STEVEN CROOK
JAMES CUNNINGHAM
NAR DAO
BEN AND SARAH DAVIS
KAREN DEAVER
SAHAN DISSANAYAKE
KEN DRAGOON
JORDAN ELLENBERG
MALCOLM FAWCETT
JOSEPHINE SAXTON FERORELLI
TODD FITCH
KATE FORESTER
MAURY FORMAN
TAYLOR GAAR
PORTIA, LUCIEN AND ARAMINTA GAITSKELL
RICHARD GAMMON
PHILIP GARLAND
(SEE PAGE 98)
SCOTT A. GEORGE
PAM GLENN
NAOMI GOLDENSON
GARETH GREEN
SONIA HAMEL
MIGEE HAN
MICHAEL HARNISH
KARL HAUSKER
JEANETTE HENDERSON
LARS HENRIKSON
THE HILDEBRAND FARM FAMILY
(IN MEMORY OF ALEX AND BARBARA)
(SEE PAGE 176)
TOM M HINCKLEY
(SEE CHIPPER, PAGE 144)
NANCY HIRSHBERG
DEBORAH C. HOARD
JOANNE HOSSACK
ELEANOR HUNGATE
NOAH ILIINSKY
TOM IMESON
THOMAS INSEL
CHANTRELLE JOHANSON
KARI JONES
VERONICA KARAS
CHANG KAWAGUCHI
MATT KELLER
ULI KINDERMANN
CHRISTOPHER W KLEIN
JOHANNA KLEIN
SUZANNA KLEIN
JIM KREMER
KAREN KURCISKA
ANDREAS LANGE
MATT LANGFORD
PATRICK LESLIE
JANE LINDLEY
VICTORIA LUGLI
(SEE KIRAN, PAGE 193)
BAS MAASE
MICHAEL MARVIN
BEN MATHEW
GUILLAUME MAUGER
ELIZABETH A MCGEE
DANIEL MELNECHUK
JONATHAN MILLER
TODD MYERS
RAMEZ NAAM
FRIKK NESJE
ANNELIESE O'LEARY
JED ODERMATT
BRANDO OMERTA
CEYDA ONER
MIEKO A. OZEKI
MARY PASCHALL
DOROTHY REILLY
(SEE PAGE 189)
KIM RIES
LEONARD RIFAS
PAUL RIPPEY
ABIGAIL ROSE
NIKOLAI ROVNEIKO
DEB RUDNICK
MONICA SAMEC AND BARRY RAWN
CHRISTIAN SARASON
BEN SCHIENDELMAN
KAI UWE SCHMIDT
JOE SHAPIRO
ERIC V SIEGEL
(AUTHOR OF *Predictive Analytics*)
SIGHTLINE INSTITUTE
RACHELLE SIMON
JOEL SINGER
TERESA MACHUSAK SMITH
AMY SNOVER
JON STAHL
ERICA STEPHAN
BETTY TANZEY
(SEE PAGE 140)
LISA TESLER
CHRIS THOUNG
JENNIFER TICE
STEPHANIE TOBOR
MARK TREXLER
(SEE PAGE 57)
DAN TRUEMAN
AXEL UTZ
SARA VIKSTRÖM
JAN-WILLEM VAN DE VEN
JOHN VECHEY
(SEE PAGE 187)
RANDALL WALL
PETER WALLACE
SPENCER WEART
(AUTHOR OF *The Discovery of Global Warming*)
MANDIE WEINANDT
DANIEL WEISE
JOHN WHITEHEAD AND TIM HAAB
(OF http://www.env-econ.net/)
(SEE PAGE 106)
KENT WHITING
JUNE WILLIAMSON
MATTHEW WILLNER
MARK WILSON
KIEL WINCH
EMILY WINTER
EDWARD WOLF
SHANNAN WONG
SPENCER WRIGHT
SAMMIE G. YOUNG, SR
ERIK ZEMLJIC
ANDREW ZWICKER

THE CARTOON INTRODUCTION TO CLIMATE CHANGE

BY GRADY KLEIN AND
YORAM BAUMAN, Ph.D.
THE WORLD'S FIRST AND ONLY STAND-UP ECONOMIST

ISLAND PRESS IS A TRADEMARK OF ISLAND PRESS/CENTER FOR RESOURCE ECONOMICS.

LIBRARY OF CONGRESS CONTROL NUMBER: 2014938166

PRINTED ON RECYCLED, ACID-FREE PAPER

MANUFACTURED IN THE UNITED STATES OF AMERICA
10 9 8 7 6 5 4 3 2 1

KEYWORDS: GLOBAL WARMING, GREENHOUSE GASES, FOSSIL FUELS, ENERGY EFFICIENCY, ECONOMIC DEVELOPMENT, CARBON DIOXIDE, ICE-AGE, MILANKOVITCH CYCLES, CAP-AND-TRADE, CARBON TAX

FOR ANNE AND LIAM AND BENJAMIN
—GK

FOR MY FRIENDS AND MENTORS AT REED COLLEGE AND SIGHTLINE INSTITUTE,
ESPECIALLY NOEL NETUSIL, MIKE FOLEY, AND ALAN DURNING
—YB

CONTENTS

Part One:
OBSERVATIONS

Part Two:
PREDICTIONS

Part Three:
ACTIONS

DATA SOURCES AND DETAILED PAGE NOTES FOR THE ENTIRE BOOK ARE AVAILABLE AT **ISLANDPRESS.ORG/CARTOON-INTRO**

PART ONE
OBSERVATIONS

CHAPTER 1
INTRODUCTION

TWO STORIES ARE *GOING TO DOMINATE* THE 21ST CENTURY.

STORY #1 IS ABOUT **ECONOMIC GROWTH,** ESPECIALLY IN POOR COUNTRIES IN **ASIA** AND **AFRICA.**

CAPITALISM AND **FREE-MARKET ECONOMICS** ARE GOING TO CREATE A LOT OF **NEW WEALTH...**

YOU'RE FEELING THE BENEFITS OF WHAT ADAM SMITH CALLED THE INVISIBLE HAND.

YOU CAN LEARN MORE ABOUT THAT IN THE CARTOON INTRODUCTION TO ECONOMICS.
THE CARTOON INTRODUCTION TO ECONOMICS
THE CARTOON INTRODUCTION TO ECONOMICS

...AND GIVE MANY MORE PEOPLE THE OPPORTUNITY TO PURSUE THEIR DREAMS.

I WANT TO BE A DANCER!
I WANT TO BE AN ASTRONAUT!

AS FAMILIES GET WEALTHIER, THEY TEND TO HAVE **FEWER CHILDREN**...

HAVING MORE KIDS IS **NOT** ONE OF MY DREAMS!
I WONDER **WHY?**

...SO THE WORLD POPULATION IS LIKELY TO **PEAK** AT ABOUT **10 BILLION** PEOPLE AND THEN SLOWLY DECLINE.

ENOUGH ALREADY!
1900
2000
2100

AS A RESULT, STORY #1 POINTS IN A DIRECTION THAT'S NOTHING SHORT OF **MIRACULOUS.**

A WORLD OF **2–6 BILLION WELL-EDUCATED** AND THEREFORE **HEALTHY** AND **WEALTHY** PEOPLE!

IT SEEMS **TOO GOOD TO BE TRUE!**

BUT WHAT ABOUT **STORY #2?**

STORY #2 IS ABOUT THE **ENVIRONMENTAL IMPACT** OF ALL THIS GROWTH AND DEVELOPMENT.
WHAT'S GOING TO HAPPEN WHEN **BILLIONS OF ASIANS** AND **AFRICANS** ALL TRY TO LIVE **LIKE AMERICANS?**
THE INVISIBLE HAND OF FREE-MARKET ECONOMICS ISN'T LIKELY TO FIX **THIS...**
...NOT WITHOUT SOME **HELP!**

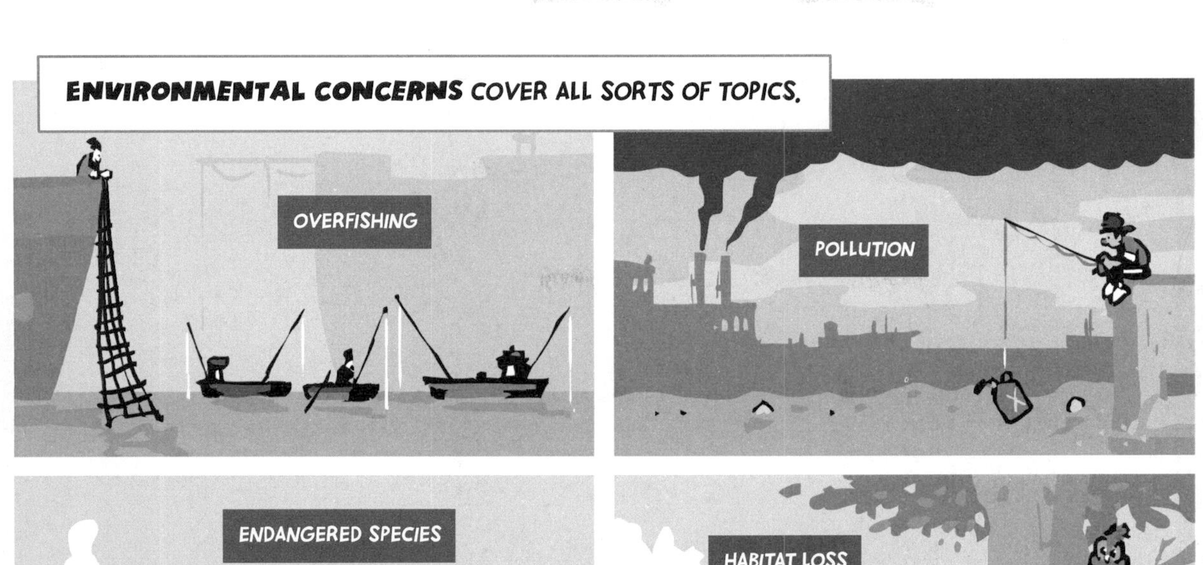
ENVIRONMENTAL CONCERNS COVER ALL SORTS OF TOPICS.
OVERFISHING
POLLUTION

ENDANGERED SPECIES

HABITAT LOSS

THIS BOOK FOCUSES ON CLIMATE CHANGE.
ALSO KNOWN AS GLOBAL WARMING.

CLIMATE CHANGE IS A **POLITICALLY CHARGED ISSUE**...

YOU'RE A DENIER!
YOU'RE AN ALARMIST!

...BUT IT MIGHT BE POSSIBLE TO FIND SOME **COMMON GROUND**...

THAT COMMON GROUND BETTER BE 6 FEET UNDER BECAUSE WE'RE ALL GOING TO DIE!

NONSENSE, WE'RE ALL GOING TO LIVE!

...BY THINKING OF CLIMATE CHANGE AS A **THREAT**.

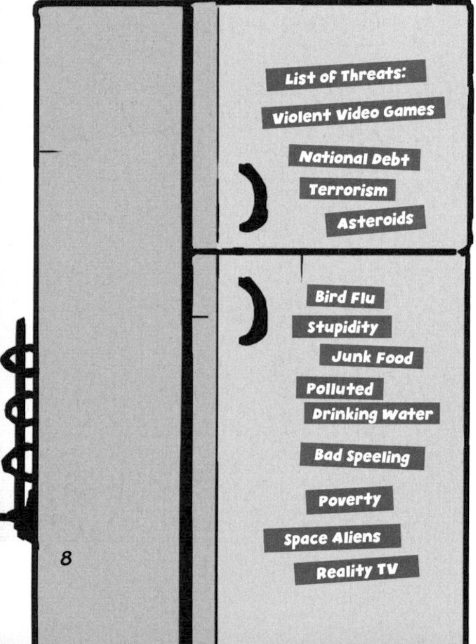
List of Threats:
Violent Video Games
National Debt
Terrorism
Asteroids
Bird Flu
Stupidity
Junk Food
Polluted Drinking Water
Bad Speeling
Poverty
Space Aliens
Reality TV

HOW DOES THIS ONE COMPARE?
Climate Change

MAYBE IT'S AN **EXISTENTIAL THREAT**...

...AND MAYBE IT'S ONLY A **MINOR THREAT**...

CLIMATE CHANGE COULD TURN THE EARTH INTO AN **ALIEN PLANET**.
HOW DO YOU KNOW IT WON'T BE AN **AWESOME** ALIEN PLANET?

...AND MAYBE IT'LL BE **DECADES** BEFORE WE KNOW FOR SURE.

AND BY THEN IT MIGHT BE **TOO LATE TO DO MORE!**
OR TO DO **LESS!**
NO WONDER CLIMATE CHANGE IS SUCH A **WICKED PROBLEM**.

THIS BOOK WILL HELP YOU **MAKE UP YOUR OWN MIND**.

AND LEARN **WHAT YOU CAN DO ABOUT IT**...
...AND WHAT WE CAN **ALL DO TOGETHER!**

YOU MIGHT FIND IT HARD TO BELIEVE THAT HUMANS CAN **INFLUENCE THE ENTIRE PLANET...**
...BUT THERE'S LOTS OF SCIENTIFIC EVIDENCE THAT **HUMAN ACTIVITY** IS **CHANGING THE CLIMATE.**
MOSTLY AS A RESULT OF **BURNING FOSSIL FUELS** AND **DEFORESTATION.**
AND SOMETIMES **BOTH AT THE SAME TIME!**
1950s
2000s

THIS BOOK IS ABOUT THE **SCIENCE** OF CLIMATE CHANGE...

CO2
CH4
H2O

...AND HOW CLIMATE CHANGE MIGHT AFFECT **LIFE ON EARTH**...

...AND **WHAT WE CAN DO ABOUT IT.**

A GOOD PLACE TO START IS WITH A FEW **DEFINITIONS**...

CLIMATE REFERS TO WHAT THE WEATHER IN A CERTAIN PLACE IS **USUALLY LIKE**.

SEATTLE IN JULY **USUALLY** HAS AFTERNOON HIGHS OF **21–28°C** (69–82°F)...

...AND IT **ALMOST NEVER RAINS!**
SOUNDS **PERFECT FOR A WEDDING!**

USUALLY DOESN'T MEAN **ALWAYS**...

...SO YOU SHOULD THINK OF CLIMATE AS **AVERAGE WEATHER**...

SORRY! THIS IS VERY **UNUSUAL**.

CLIMATE IS LIKE YOUR **PERSONALITY**.
WEATHER IS LIKE YOUR **MOOD**.

...AND **CLIMATE CHANGE** AS A **CHANGE IN AVERAGE WEATHER**.

ON AVERAGE, THESE FLOWERS NOW BLOOM **ONE WEEK EARLIER THAN THEY DID 30 YEARS AGO**.

. . . AND WITH AN **OBSERVATION.**

EARTH'S CLIMATE HAS **ALWAYS BEEN IN FLUX.**
HERE ON PLANET EARTH, CLIMATE CHANGE HAS **ALWAYS BEEN A FACT OF LIFE. . .**
. . . AND IT **ALWAYS WILL BE.**

BUT IF THE CLIMATE IS ALWAYS CHANGING THEN **WHAT'S DIFFERENT THIS TIME?**
WHAT DO **PEOPLE** HAVE TO DO WITH IT?
GOOD QUESTION!
LET'S START TO ANSWER IT BY TAKING A BRIEF LOOK AT THE **HISTORY OF PLANET EARTH.**

CHAPTER 2
A BRIEF HISTORY OF PLANET EARTH

TELL ME EVERYTHING THAT'S HAPPENED SO FAR.
WELL, FIRST THE EARTH COOLED...
...AND THEN THE DINOSAURS CAME...

THE EARTH FORMED ABOUT **4.6 BILLION** YEARS AGO.

4.6 bya | 4 bya | 3 bya | 2 bya | 1 bya | Now

THAT'S 50 MILLION TIMES OLDER THAN GRANDPA.

IN THE EARLY DAYS IT WAS A GREAT **MOLTEN BALL OF LIQUID ROCK**...

... WHICH BELCHED OUT **HOT GASES** AND **WATER VAPOR** TO FORM THE EARLY **ATMOSPHERE** AND **OCEANS**.

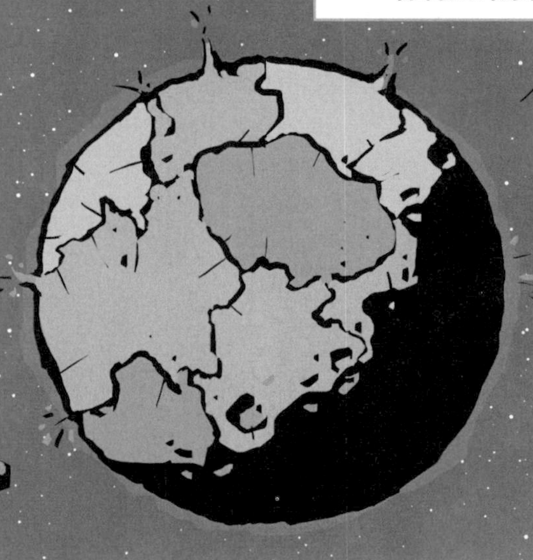

PRIMITIVE LIFE FORMS APPEARED IN THE OCEANS SOMETIME WITHIN THE FIRST **BILLION YEARS**.

4.6 bya | Now

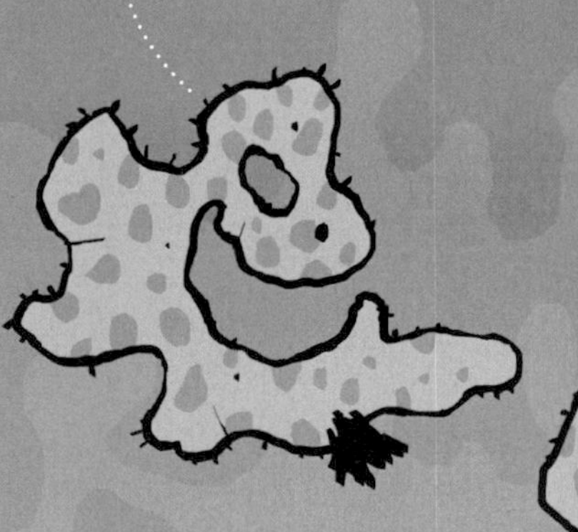
WHAT DO YOU CALL A SINGLE-CELLED ORGANISM FLOATING IN A PRIMORIDAL SEA?
BOB.

WHAT DO YOU CALL A SINGLE-CELLED ORGANISM SHAPED LIKE A TUBE?
ROD.

YOU AND YOUR JOKES ARE SO PRIMITIVE!

THEN, AFTER **HUNDREDS OF MILLIONS OF YEARS OF EVOLUTION**...

4.6 bya
Now
WHAT DO YOU CALL A SINGLE-CELLED ORGANISM THAT LIVES IN YOUR MAILBOX?
BILL.

WHAT DO YOU CALL A SINGLE-CELLED ORGANISM SMEARED ON THE WALL?
ART.

WILL YOU TWO SHUT UP!
I'M TRYING TO CONCENTRATE!

...SOME OF THESE ORGANISMS FIGURED OUT OXYGENIC **PHOTOSYNTHESIS**...

EUREKA!
EEK! HARRIET, PUT ON SOME CLOTHES!

...WHICH IS THE **CHEMICAL REACTION** THAT ALLOWS **GREEN THINGS** LIKE **PLANTS** AND **ALGAE** TO GROW.

IT TURNS SUNLIGHT, WATER, AND CARBON DIOXIDE...
...INTO THINGS LIKE BROCCOLI AND BEAN SPROUTS.

OH GREAT.
GREEN THINGS ARE THE BASE OF THE FOOD CHAIN.
WITHOUT THEM THERE COULD BE NO US.

NOWADAYS, GREEN THINGS PLAY A KEY ROLE IN THE **CARBON CYCLE...**

IN CHAPTER 4 WE'LL SEE THE INFLUENCE OF **HUMAN ACTIVITY**.

EVERY YEAR ABOUT **80 BILLION TONS** OF ATMOSPHERIC CARBON **DISSOLVES IN SEAWATER...**

...AND ABOUT THAT MUCH RETURNS TO THE ATMOSPHERE THROUGH **OUTGASSING**.

MEANWHILE, ABOUT **120 BILLION TONS** GETS SUCKED IN BY **PLANTS** THROUGH **PHOTOSYNTHESIS...**

...AND ABOUT THAT MUCH RETURNS TO THE ATMOSPHERE THROUGH **FIRE**, **DECOMPOSITION**, AND **RESPIRATION** BY PLANTS AND ANIMALS.

AND THERE'S LOTS MORE, LIKE MIXING BETWEEN THE **SURFACE OCEAN** AND THE **DEEP OCEAN**.

...WHICH IS CRUCIAL TO **ALL LIFE ON EARTH**.

YOU **ARE** WHAT YOU **EAT!**
Mostly Water and Carbon
Mostly Water and Carbon
Mostly Water and Carbon

BETWEEN ABOUT 2.8 AND 2.3 **BILLION YEARS AGO**, HOWEVER, GREEN THINGS DID SOMETHING PERHAPS EVEN **MORE IMPORTANT**.

4.6 bya
Now

GREEN SLIME IS GROSSLY UNDERAPPRECIATED.
THEY PUMPED LOTS AND LOTS OF OXYGEN INTO THE ATMOSPHERE.

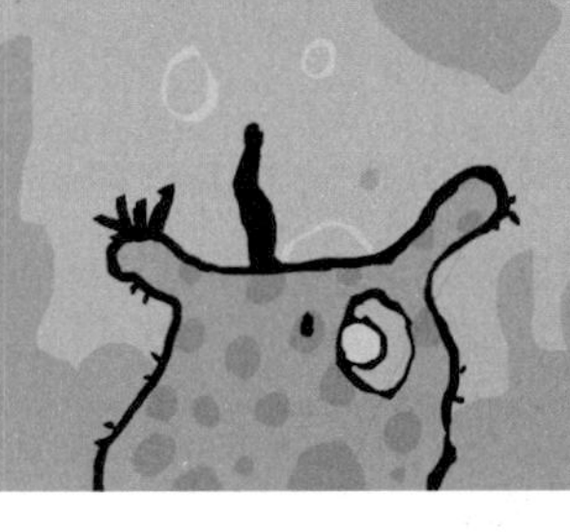

GETTING **OXYGEN** (O_2) INTO THE **ATMOSPHERE** WAS **REALLY IMPORTANT** BECAUSE...

DUH! BECAUSE ANIMALS NEED OXYGEN TO BREATHE.

WELL, THAT'S TRUE...
...BUT THIS WAS A BILLION YEARS BEFORE ANIMALS.

...BECAUSE IT LED TO THE CREATION OF A **LAYER OF OZONE** (O_3) ABOUT 15 MILES ABOVE THE SURFACE OF THE PLANET.

WHAT'S SO IMPORTANT ABOUT THAT?
WELL, DUH.
TURN THE PAGE TO FIND OUT.
OZONE LAYER

BEFORE THE OZONE LAYER,
LIFE HAD TO **HIDE FROM THE SUN.**

4.6 bya — Now

THE SUN GENERATES **VISIBLE LIGHT...**
...BUT IT ALSO GENERATES **DEADLY ULTRAVIOLET RADIATION.**
THAT'S WHY WE LIVE **UNDERWATER** OR **UNDER ROCKS.**

AFTER THE OZONE LAYER, THE
SUN'S **DEADLIEST UV RAYS** GOT **BLOCKED...**

4.6 bya — Now

OZONE IS LIKE **SUNSCREEN FOR THE ENTIRE PLANET.**

Sunscreen
Protects against
UVA & UVB
For up to 2 Hours!

Ozone
Protects against
DEADLY UVC
For Billions of Years.
Now with Catalytic Action!
3

...AND THAT ALLOWED LIFE TO MOVE **INTO THE SUNSHINE.**

EUREKA!
NO, HARRIET, **NOT AGAIN!**

NO WONDER EVERYBODY GOT WORRIED WHEN SCIENTISTS DISCOVERED
A HOLE IN THE OZONE LAYER IN THE 1980s.

DESPITE WHAT MANY PEOPLE THINK, THE OZONE HOLE IS **NOT** CLOSELY RELATED TO GLOBAL WARMING...

ENVIRONMENTAL PROBLEMS ARE **NOT** ALL THE SAME...
...AND YOU CAN'T SOLVE THEM ALL BY **RECYCLING**.

...BUT IT IS VALUABLE TO **COMPARE** AND **CONTRAST** THE TWO ISSUES...

THE OZONE HOLE IS RELATED TO HUMAN EMISSIONS OF **OZONE-DESTROYING** GASES SUCH AS **CHLOROFLUOROCARBONS (CFCs)**.

GLOBAL WARMING IS RELATED TO HUMAN EMISSIONS OF **GREENHOUSE** GASES SUCH AS **CARBON DIOXIDE (CO_2)**.

...AND WE CAN TAKE HEART FROM THE **PROGRESS** WE'VE MADE IN REPAIRING THE OZONE HOLE.

PRESIDENT REAGAN SIGNED THE **MONTREAL PROTOCOL** THAT HELPED PHASE OUT CFCs...
...AND THE OZONE LAYER IS **RECOVERING**.

IF ONLY GLOBAL WARMING WERE SO **EASY**!

BUT LET'S GET BACK TO THE **HISTORY OF PLANET EARTH**...

THE **PAST BILLION YEARS** HAVE SEEN **WILD CHANGES** IN THE **CLIMATE.**
4.6 bya
Now
THE ONLY THING **PERMANENT** IS **CHANGE.**

THERE WERE TIMES WHEN **ICE COVERED ALMOST EVERYTHING...**

THIS **SNOWBALL EARTH** IS FREEZING!
GOOD THING THERE WEREN'T ACTUALLY **PEOPLE** BACK THEN.

...AND TIMES WHEN **THE NORTH POLE WAS TROPICAL.**

BOY, I COULD REALLY USE SOME **ICED TEA!**
SORRY, THERE'S **NO ICE.**
GOOD THING THERE WEREN'T ACTUALLY **PEOPLE** BACK THEN.

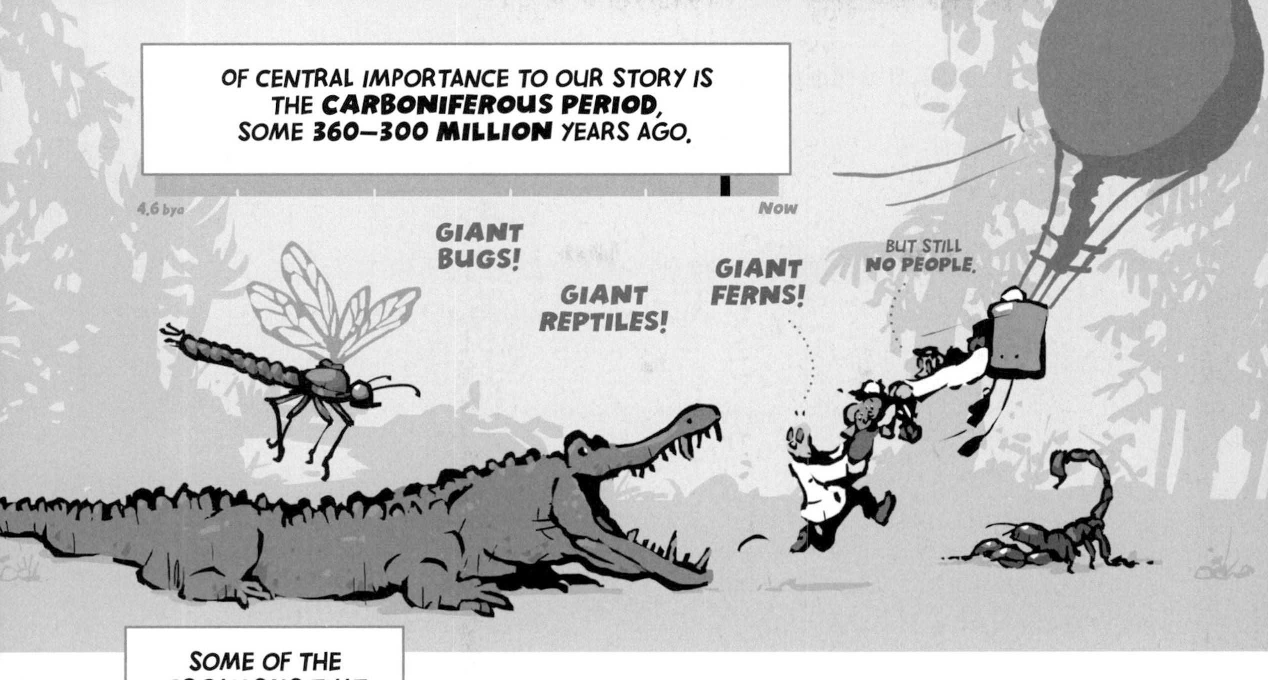
OF CENTRAL IMPORTANCE TO OUR STORY IS THE **CARBONIFEROUS PERIOD**, SOME **360–300 MILLION** YEARS AGO.
4.6 bya
Now
GIANT BUGS!
GIANT REPTILES!
GIANT FERNS!
BUT STILL NO PEOPLE.

SOME OF THE ORGANISMS THAT **DIED** THEN GOT BURIED...
WHAT DO YOU CALL A SINGLE-CELLED ORGANISM **LYING IN A HOLE?**
DOUG.

...AND COOKED UNDERGROUND FOR **HUNDREDS OF MILLIONS OF YEARS**...

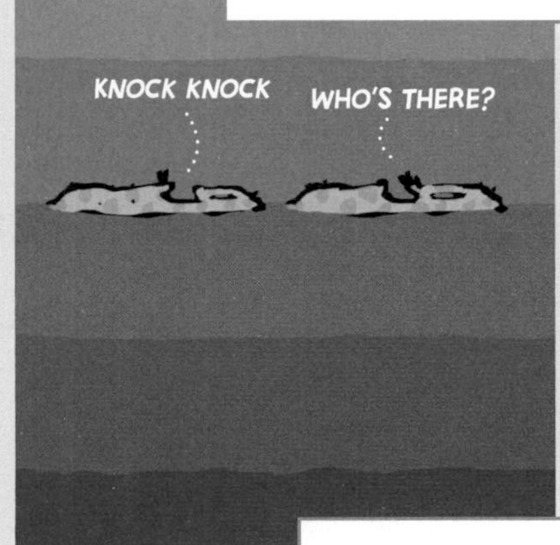
KNOCK KNOCK
WHO'S THERE?

OIL.
OIL **WHO?**

OIL BE BACK.

...AND EVENTUALLY TURNED INTO **CARBON-BASED FOSSIL FUELS.**

THAT'S WHERE A LOT OF OUR **COAL** COMES FROM...

...AND PLENTY OF **OIL** AND **NATURAL GAS**, TOO.

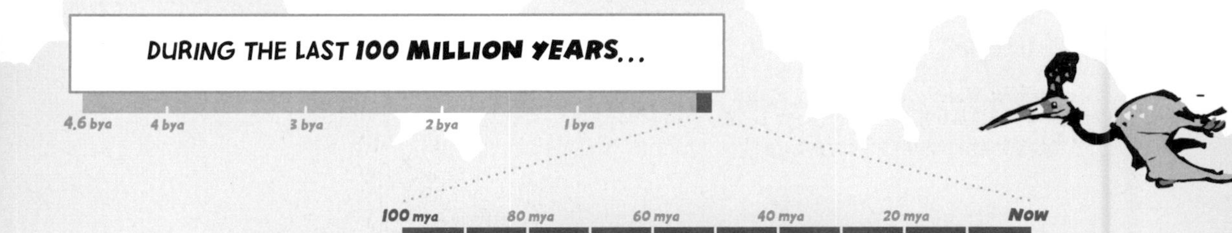
DURING THE LAST **100 MILLION YEARS...**
4.6 bya
4 bya
3 bya
2 bya
1 bya
100 mya
80 mya
60 mya
40 mya
20 mya
Now

...THE CONTINENTS **SLOWLY DRIFTED** INTO THE POSITIONS THEY HAVE TODAY...

CONTINENTS MOVE AT ABOUT THE SPEED THAT YOUR **FINGERNAILS GROW...**
...ABOUT **100 MILES** EVERY **MILLION YEARS.**

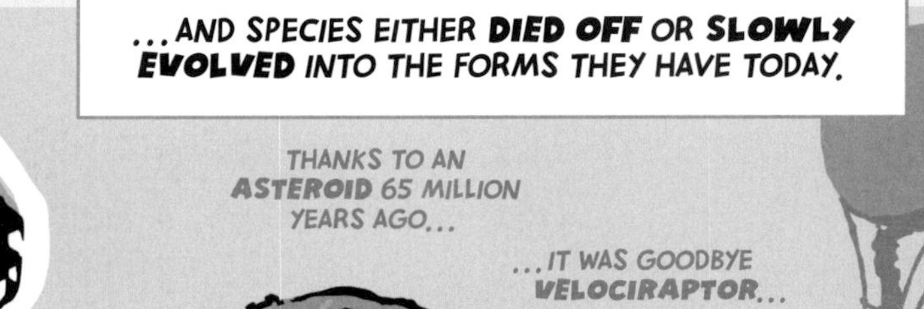
...AND SPECIES EITHER **DIED OFF** OR **SLOWLY EVOLVED** INTO THE FORMS THEY HAVE TODAY.

THANKS TO AN **ASTEROID** 65 MILLION YEARS AGO...

...IT WAS GOODBYE **VELOCIRAPTOR...**
...AND HELLO **CHICKEN.**

100 mya
Now

EVEN MORE RECENTLY, THE EARTH'S CLIMATE HAS **CALMED DOWN A BIT...**

30 MILLION YEARS WITH NO **SNOWBALL EARTHS...**
...AND NO **TROPICAL NORTH POLES!**

...BUT **CALM** IS A **RELATIVE TERM.**

LOOK, WE'VE **CALMED DOWN!**
WOW, WHAT WERE YOU LIKE **BEFORE?**

100 mya
Now

IN PARTICULAR, OVER THE PAST **2.6 MILLION YEARS** THE PLANET HAS GONE THROUGH **REPEATED CYCLES**...

...OF **WARM PERIODS**...

...AND **COOL PERIODS.**

SCIENTISTS CALL THEM **GLACIAL PERIODS.**
EVERYONE ELSE CALLS THEM **ICE AGES.**

100 mya — Now

FINALLY, ABOUT **200,000 YEARS AGO...**

THAT'S ONLY 2,000 TIMES OLDER THAN GRANDPA.

A BLINK OF AN EYE IN THE EARTH'S LIFESPAN.

...ANATOMICALLY MODERN **HUMAN BEINGS** APPEARED IN AFRICA.

TA DA!

IT WASN'T LONG BEFORE THEY STARTED ASKING **TOUGH QUESTIONS.**

I WONDER HOW I CAN AVOID BEING EATEN?
I WONDER WHERE I CAN FIND SOME HOT SAUCE?
I WONDER WHAT CAUSED THE ICE AGES?

CHAPTER 3
THE ICE AGES

GO BACK 15,000 YEARS
AND THIS WAS UNDER
A **MILE OF ICE!**

EARLY GEOLOGISTS HYPOTHESIZED THAT THERE **MUST** HAVE BEEN **ICE AGES.**
WHAT COULD HAVE CARRIED **THESE GIANT ROCKS** HERE FROM **THOSE DISTANT MOUNTAINS?**
THERE MUST HAVE BEEN **MASSIVE GLACIERS!**
OR **PREHISTORIC GIANTS!**

20TH CENTURY SCIENTISTS **CONFIRMED** THIS BY STUDYING EARTH'S TWO REMAINING **GIANT ICE SHEETS.**

ONE'S OVER **ANTARCTICA...**
...THE OTHER'S OVER **GREENLAND.**

THESE ICE SHEETS WERE BUILT UP OVER **HUNDREDS OF THOUSANDS OF YEARS.**

FALLING SNOW GOT
COMPRESSED,
TRAPPING ANCIENT AIR IN ANCIENT ICE.

BY **DRILLING DOWN THROUGH THAT ICE...**

I'VE GOT A 6–FOOT **COD!**

I'VE GOT A MILE–LONG **ICE CORE!**

...AND **ANALYZING** THE VARIOUS LAYERS...

ICE CORES ARE A BIT LIKE **TREE RINGS.**

...SCIENTISTS CAN ESTIMATE THE **AVERAGE SURFACE TEMPERATURE** OF PLANET EARTH FAR INTO THE **DISTANT PAST.**

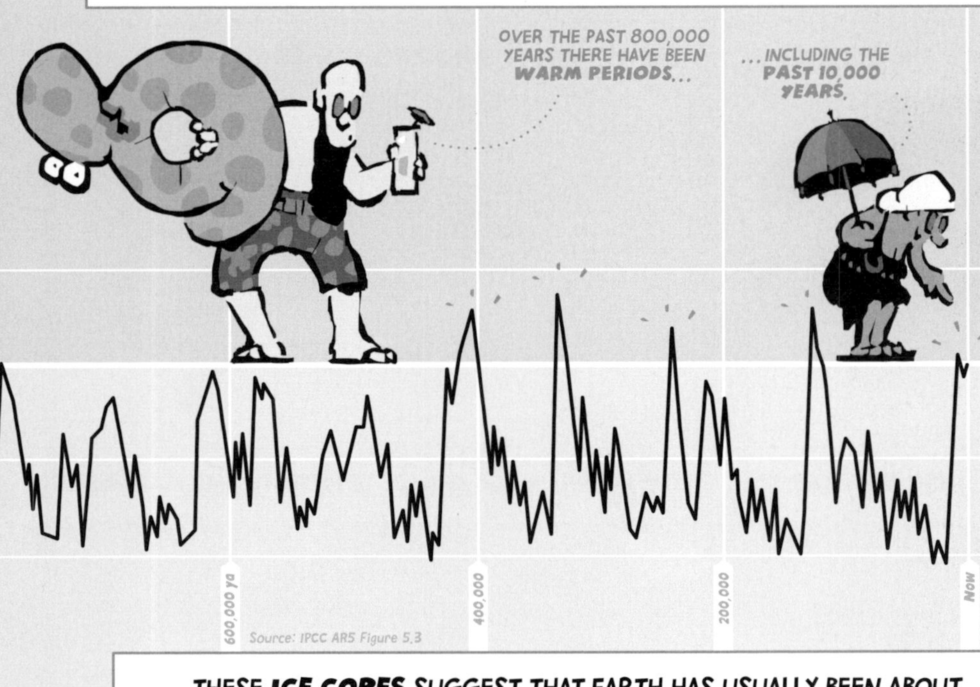
OVER THE PAST 800,000 YEARS THERE HAVE BEEN **WARM PERIODS...**
...INCLUDING THE **PAST 10,000 YEARS.**
+4°C
Average Today
-4°C
-8°C
BUT THOSE WARM PERIODS HAVE BEEN THE **EXCEPTIONS.**
600,000 ya
400,000
200,000
Now
Source: IPCC AR5 Figure 5.3

THESE **ICE CORES** SUGGEST THAT EARTH HAS USUALLY BEEN ABOUT **6°C** (11°F) **COLDER** THAN IT IS NOW.

THAT DIFFERENCE IN TEMPERATURE MAY **SEEM SMALL...**
...BUT IT WAS BIG ENOUGH FOR GLACIERS TO **BURY CHICAGO.**

THE MYSTERY OF THE ICE AGES...
I WONDER WHY THERE AREN'T GLACIERS HERE ANYMORE?
AND DON'T SAY PREHISTORIC GIANTS!
...WAS MOSTLY SOLVED BY THE SERBIAN MATHEMATICIAN MILUTIN MILANKOVITCH DURING THE FIRST WORLD WAR.

COMPARED TO THE MYSTERY OF WHY WE'RE FIGHTING THIS DUMB WAR...
...THIS IS CHILD'S PLAY.
HE STUDIED THE DETAILS OF THE EARTH'S ORBIT AROUND THE SUN...

IT ORBITS ONCE A YEAR...
...BUT OVER THE YEARS IT MEANDERS AND WOBBLES A BIT.
...INCLUDING THE TILT OF THE EARTH, WHICH CAUSES WINTERS AND SUMMERS.
IN JULY THE NORTHERN HEMISPHERE GETS MORE SUN...
...AND THE SOUTHERN HEMISPHERE IS DARK AND COLD.
April
October
IN JANUARY THE NORTHERN HEMISPHERE IS DARK AND COLD...
...AND THE SOUTHERN HEMISPHERE GETS MORE SUN.
NOT TO SCALE.

IT TURNS OUT THAT THE TILT **VARIES** OVER TENS OF THOUSANDS OF YEARS.

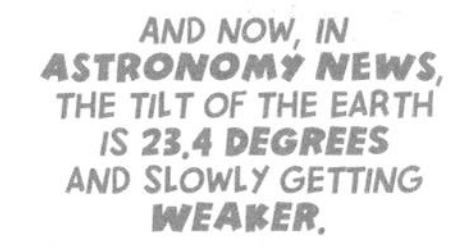
AND NOW, IN **ASTRONOMY NEWS**, THE TILT OF THE EARTH IS **23.4 DEGREES** AND SLOWLY GETTING **WEAKER**.

WE'LL HAVE AN UPDATE IN THE YEAR **11,000**.

WHEN THE TILT IS STRONGER, **SEASONS ARE STRONGER**...
MY HEAD IS **ON FIRE**...
...AND MY FEET ARE **FREEZING!**
MY **HEAD** IS **FREEZING**...
...AND MY **FEET** ARE **ON FIRE!**

...AND WHEN THE TILT IS WEAKER, **SEASONS ARE WEAKER**.
I'M GETTING COOKED A MORE **EVEN AMOUNT ALL AROUND**...
...NO MATTER WHAT TIME OF YEAR IT IS.

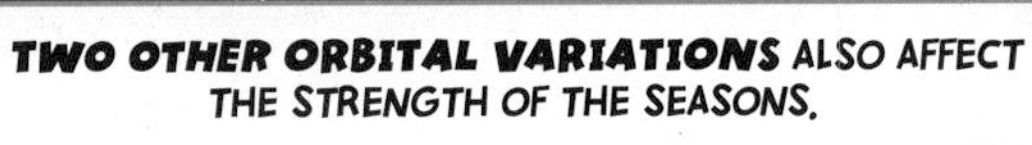
TWO OTHER ORBITAL VARIATIONS ALSO AFFECT THE STRENGTH OF THE SEASONS.

IN ADDITION TO **TILT**, THERE'S **ECCENTRICITY** AND **PRECESSION**...
...BUT YOU CAN THINK OF THEM AS **WIGGLES** AND **JIGGLES**.

THE **MILANKOVITCH CYCLE** THEORY SAYS THAT THESE THREE ORBITAL VARIATIONS...

...SET THE **RHYTHM** FOR THE ICE AGES...

LIKE A **PACEMAKER.**

...BY TRIGGERING **POSITIVE FEEDBACK LOOPS.**

SODA MAKES ME **GO CRAZY**...
GOING CRAZY MAKES ME **THIRSTY**...

POSITIVE FEEDBACK LOOPS **AMPLIFY CHANGES**...
...SO THAT **CRAZY** TURNS INTO **EVEN MORE CRAZY.**

SEE THE **GLOSSARY** FOR DETAILS.

SOMETIMES THOSE POSITIVE FEEDBACK LOOPS BRING THE EARTH **OUT OF AN ICE AGE**...

I'M GOING **INTERGLACIAL!**

...AND SOMETIMES THEY SEND US **BACK INTO ONE.**

GETTING **SLEEPY**...
...MAKES ME **EVEN MORE SLEEPY.**
ZZZZZZZZ.

LET'S SEE HOW IT WORKS IN **CLOSER DETAIL.**

FOR A **SIMPLIFIED VERSION** OF THE MILANKOVITCH STORY...

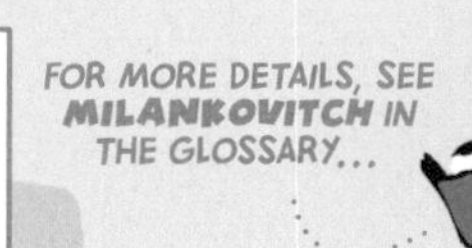
FOR MORE DETAILS, SEE **MILANKOVITCH** IN THE GLOSSARY...

...OR READ MY **626-PAGE BOOK!**

...IMAGINE WE'RE IN **CANADA** DURING A **GLACIAL PERIOD.**

LIKE THE ONE THAT PEAKED **20,000 YEARS AGO.**

THE MILANKOVITCH CYCLES EVENTUALLY CREATE CONDITIONS WITH **STRONG SEASONS.**

HOT HOT SUMMERS.

LOTS OF **MELTING ICE AND SNOW.**

COLD COLD WINTERS.

BUT THERE ISN'T ENOUGH **NEW SNOW** TO KEEP UP WITH THE **SUMMER MELT.**

THAT CAUSES **ICE AND SNOW** TO SLOWLY GIVE WAY TO **LAND AND WATER.**

NOW WE CAN PLANT **MAPLE SYRUP TREES!**

AMPLIFYING THE MILANKOVITCH CYCLES ARE THINGS LIKE THE **ICE–ALBEDO EFFECT...**

ICE AND SNOW **REFLECT** LOTS OF SUNLIGHT BACK INTO SPACE.
LAND AND WATER **ABSORB** MORE SUNLIGHT.

...WHICH CREATE **POSITIVE FEEDBACK LOOPS...**

MELTING ICE AND SNOW LEADS TO **HIGHER TEMPERATURES...**

HIGHER TEMPERATURES **MELT ICE AND SNOW...**

WARMING TURNS INTO **EVEN MORE WARMING.**

...THAT BRING ABOUT A GLOBAL **INTERGLACIAL PERIOD.**

IT SURE IS **HOT** AND **LOUD** IN HERE!

WHAT?

TO SEE HOW IT WORKS IN THE OTHER DIRECTION, IMAGINE WE'RE IN THE MIDDLE OF AN **INTERGLACIAL PERIOD.**

LIKE THE ONE WE'RE IN **NOW.**
The Mammoth Room
TONITE 7 pm, No Cover
Milankovitch and the Feedbacks

THE MILANKOVITCH CYCLES EVENTUALLY CREATE CONDITIONS WITH **MILD SEASONS.**

MILD SUMMERS.

NOT MUCH MELTING ICE AND SNOW.

MILD WINTERS.

BUT STILL COLD ENOUGH FOR SNOW IN PLACES LIKE **CANADA.**

THAT CAUSES **ICE AND SNOW** TO SLOWLY REPLACE **LAND AND WATER.**

WHAT'S MORE CANADIAN THAN **MAPLE SYRUP?**
ICE HOCKEY!

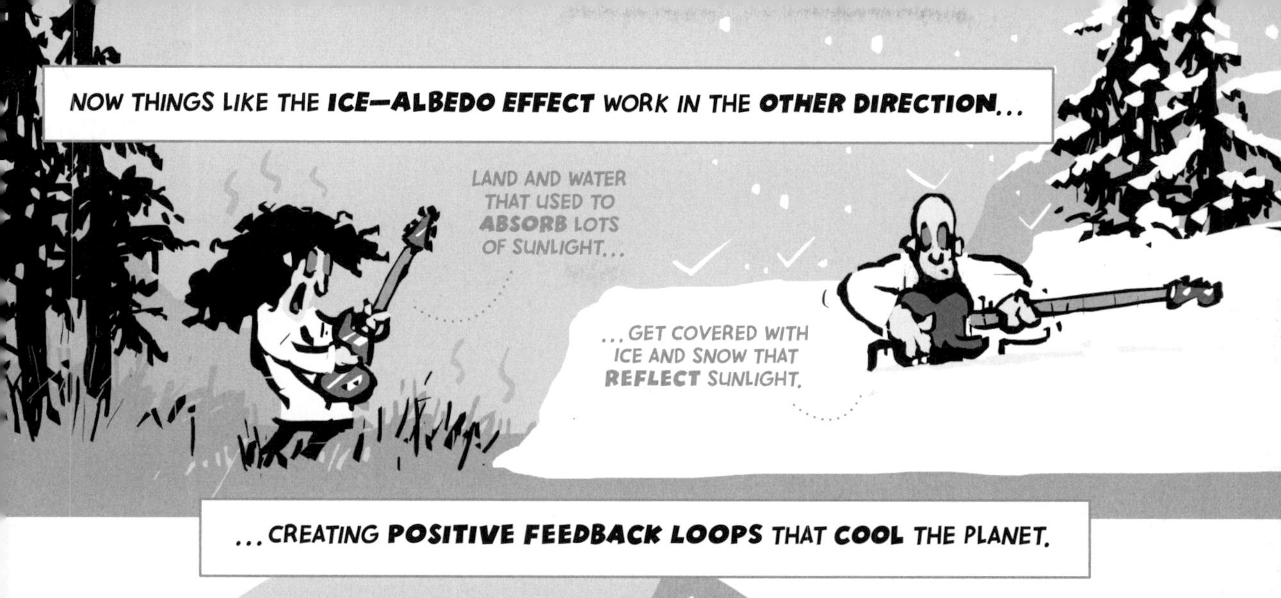
NOW THINGS LIKE THE **ICE—ALBEDO EFFECT** WORK IN THE **OTHER DIRECTION**...
LAND AND WATER THAT USED TO **ABSORB** LOTS OF SUNLIGHT...
...GET COVERED WITH ICE AND SNOW THAT **REFLECT** SUNLIGHT.

...CREATING **POSITIVE FEEDBACK LOOPS** THAT **COOL** THE PLANET.

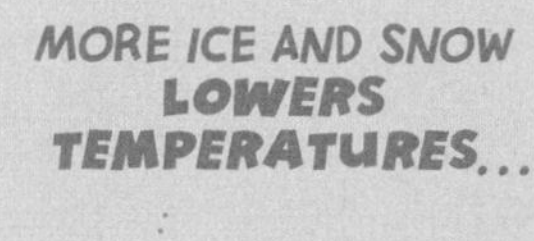
MORE ICE AND SNOW **LOWERS TEMPERATURES**...

LOWER TEMPERATURES MEAN **MORE ICE AND SNOW**...

COOLING TURNS INTO **EVEN MORE COOLING**.

ALL THIS BRINGS UP A **PRETTY OBVIOUS QUESTION**.

GIVEN THAT WE'RE **CURRENTLY** IN AN **INTERGLACIAL PERIOD**...
...SHOULD WE WORRY ABOUT THE **NEXT ICE AGE?**

THE **ANSWER** IS THAT MOTHER NATURE **IS** GEARING UP FOR **ANOTHER ICE AGE...**

OH NO, WE BETTER **STOCK UP ON HEATING OIL!**

...BUT NOT FOR AT LEAST **30,000 YEARS.**

HA!
NEVER MIND!

AND LONG BEFORE **THAT** HAPPENS WE'RE GOING TO HAVE TO COME TO TERMS WITH WHAT **WE'RE** DOING TO THE CLIMATE.

UM, ABOUT THAT **HEATING OIL...**

CHAPTER 4
CARBON DIOXIDE

EMPTY YOUR MIND . . .
. . . AND FOCUS ON SOMETHING **ODORLESS**, **COLORLESS**, AND **INVISIBLE**.

WHAT WE CALL **AIR** IS ACTUALLY A MIXTURE THAT'S ABOUT **21% OXYGEN**...
...AND ABOUT **78% NITROGEN**.
CREATED OVER THE EONS BY ALL THAT PHOTOSYNTHESIZING **GREEN STUFF**.
7th Annual Nitrogen Day
N_2 DEEP
I'm into N_2
N_2 the Nth Power
N_2 the Dragon
THE **REMAINING 1%** INCLUDES **WATER VAPOR**...

...AND RELATIVELY **TINY AMOUNTS** OF OTHER GASES...

...INCLUDING **CARBON DIOXIDE (CO_2)**.

AIR IS ONLY ABOUT 0.04% CO_2...
...BUT THAT STILL EQUALS ABOUT 4,000,000,000,000,000,000 MOLECULES OF CO_2 IN EVERY BREATH YOU TAKE.

IN THE EARLY 1950s A CHEMIST NAMED **CHARLES DAVID KEELING...**

CALL ME **DAVE.**

...FIGURED OUT A WAY TO ACCURATELY MEASURE **THE CONCENTRATION OF CARBON DIOXIDE IN AIR.**

WHAT'S THE CO_2 DIFFERENCE BETWEEN THESE TWO SAMPLES?
ONE IN A MILLION!

STARTING IN 1958, KEELING AND HIS COLLEAGUES MADE **DAILY MEASUREMENTS** OF CO_2 AT THE **MAUNA LOA OBSERVATORY** IN **HAWAII...**

C'MON DAD, LET'S GO TO THE BEACH!
SORRY, KID, NOT NOW.

...AND THE RESULTS MADE HIM **FAMOUS.**

KEELING MADE **TWO BIG DISCOVERIES.** FIRST, HE DISCOVERED AN **ANNUAL CYCLE** IN CO_2 CONCENTRATIONS.

CO_2 concentrations [parts per million]
325
320
315
310
NOV 1959
MAY 1960
NOV 1960
MAY 1961
NOV 1961
DUDE, IT'S LIKE THE PLANET IS **BREATHING!**
OKAY, HIPPIE, TAKE IT EASY.
Source: http://scrippsCO2.ucsd.edu

THIS CYCLE IS RELATED TO THE **SEASONS...**

LOOK AT PAGE 30!

...AND TO THE **CARBON CYCLE...**

LOOK AT PAGE 18!
IN
OUT

...AND TO THE FACT THAT **MOST OF THE LAND** ON EARTH IS IN THE **NORTHERN HEMISPHERE.**

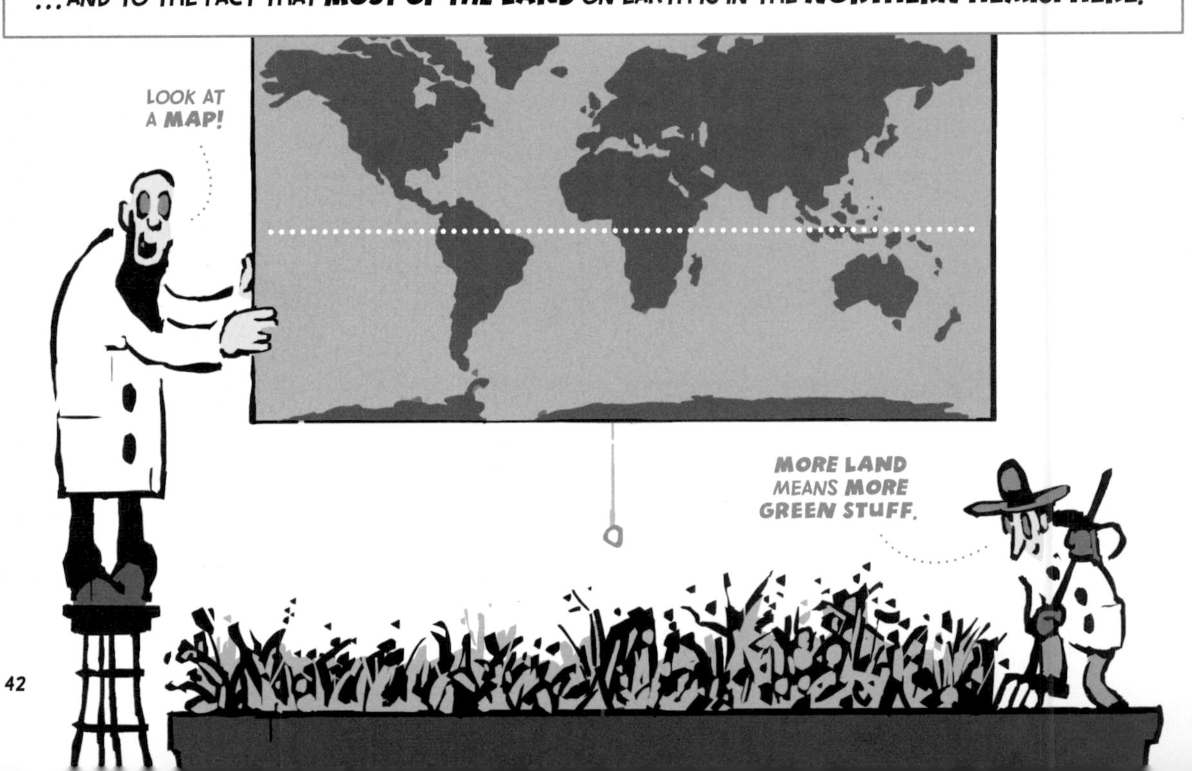

LOOK AT A **MAP!**
MORE LAND MEANS **MORE GREEN STUFF.**

WHEN IT'S **WINTER** IN THE NORTH,
EXTRA CO_2 GETS "EXHALED" INTO THE ATMOSPHERE...

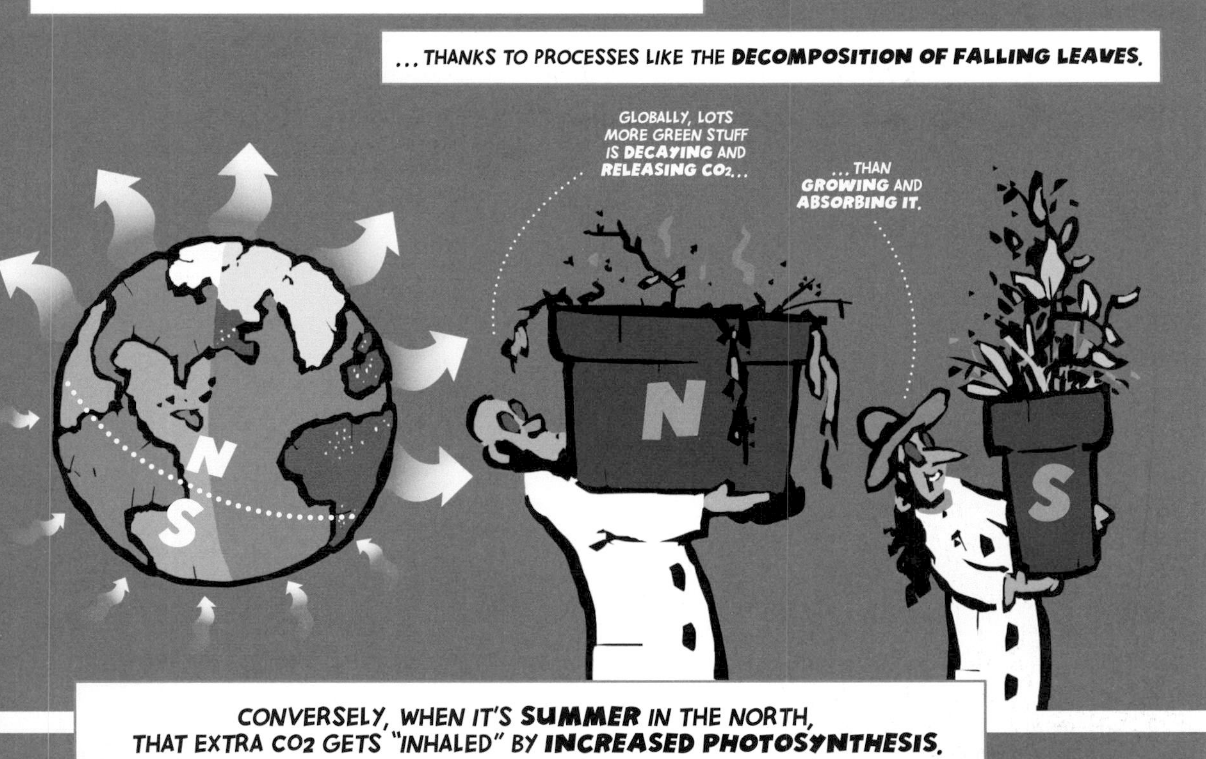
...THANKS TO PROCESSES LIKE THE **DECOMPOSITION OF FALLING LEAVES.**
GLOBALLY, LOTS MORE GREEN STUFF IS **DECAYING** AND **RELEASING CO_2...**
...THAN **GROWING** AND **ABSORBING IT.**
N
S
N
S

CONVERSELY, WHEN IT'S **SUMMER** IN THE NORTH,
THAT EXTRA CO_2 GETS "INHALED" BY **INCREASED PHOTOSYNTHESIS.**

GLOBALLY, LOTS MORE GREEN STUFF IS **GROWING** AND **ABSORBING CO_2...**
...THAN **DECAYING** AND **RELEASING IT.**
N
S
N
S

KEELING'S **SECOND BIG DISCOVERY** WAS THAT CO_2 CONCENTRATIONS IN THE ATMOSPHERE WERE **INCREASING OVER TIME.**

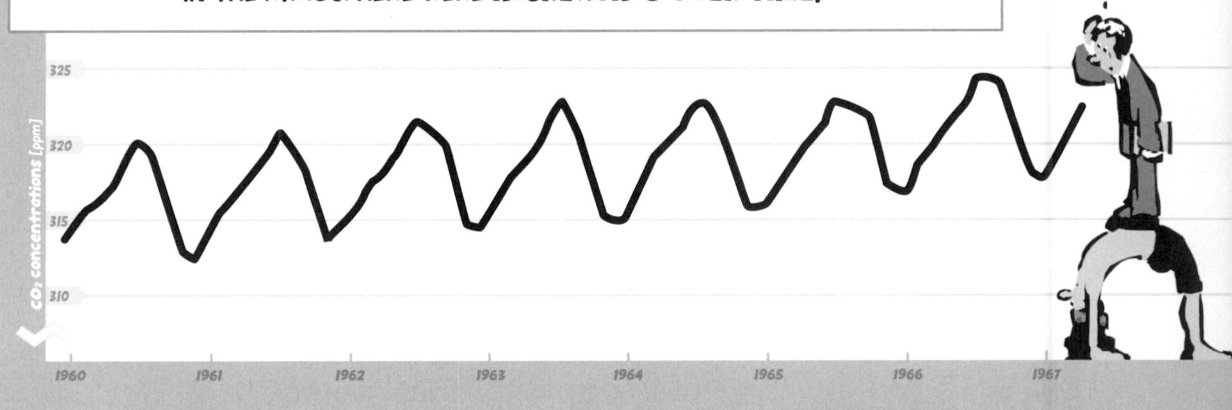

CO_2 concentrations [ppm]
325
320
315
310
1960
1961
1962
1963
1964
1965
1966
1967
Source: http://scrippsCO2.ucsd.edu

WHEN KEELING STARTED IN 1958, THE INCREASE WAS ABOUT **1 PART PER MILLION (ppm) PER YEAR.**

AN INCREASE OF 1 ppm EQUALS ABOUT 2 BILLION EXTRA TONS OF CARBON...
...OR ALMOST 8 BILLION EXTRA TONS OF CO2.
SEE CARBON IN THE GLOSSARY FOR DETAILS.

KEELING'S MEASUREMENTS CONTINUED FOR **YEARS**, AND THEN **DECADES**, AND AFTER HIS RETIREMENT HIS SON STEPPED INTO HIS SHOES.

C'MON SON, LET'S GO TO THE BEACH!
SORRY, DAD, NOT NOW.

THE GRAPH OF THE DAILY MEASUREMENTS THAT THEY AND THEIR COLLEAGUES HAVE MADE SINCE 1958 IS CALLED THE **KEELING CURVE.**

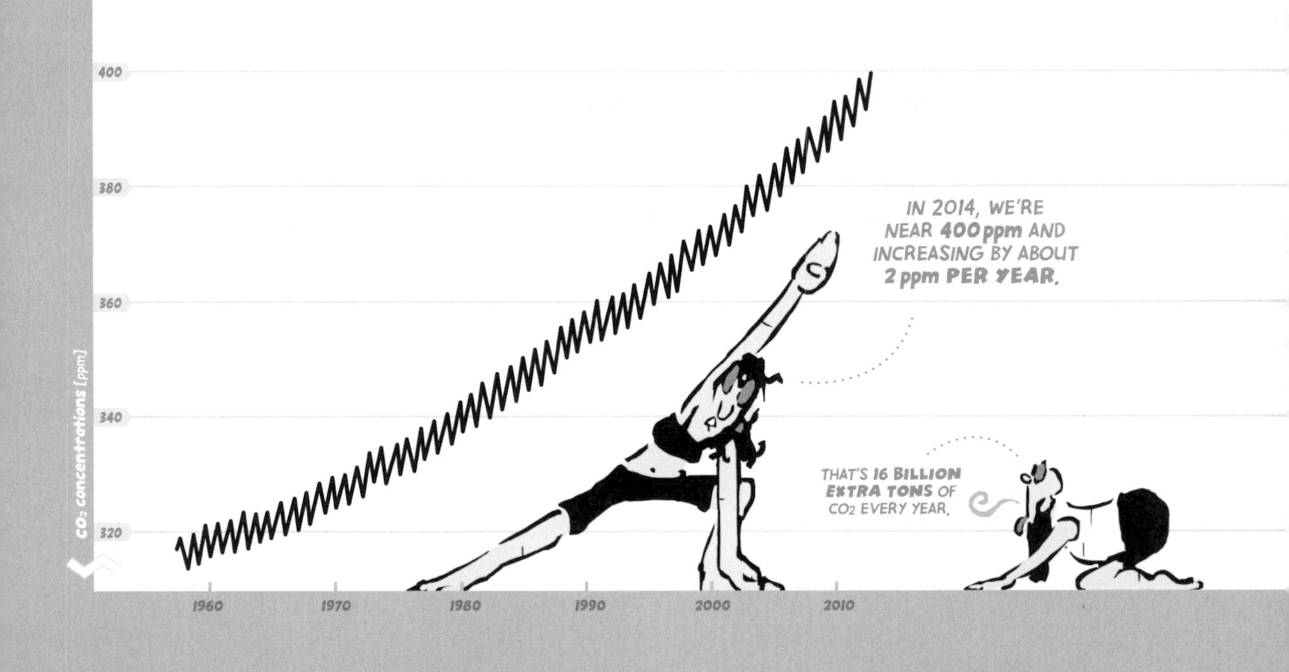

400
380
360
340
320
CO_2 concentrations [ppm]
1960
1970
1980
1990
2000
2010
IN 2014, WE'RE NEAR **400 ppm** AND INCREASING BY ABOUT **2 ppm PER YEAR.**
THAT'S **16 BILLION EXTRA TONS** OF CO_2 EVERY YEAR.
Source: http://scrippsco2.ucsd.edu

IT IS NOW ONE OF THE **MOST FAMOUS IMAGES** IN THE WORLD...

Campbell's
CONDENSED
TOMATO
SOUP

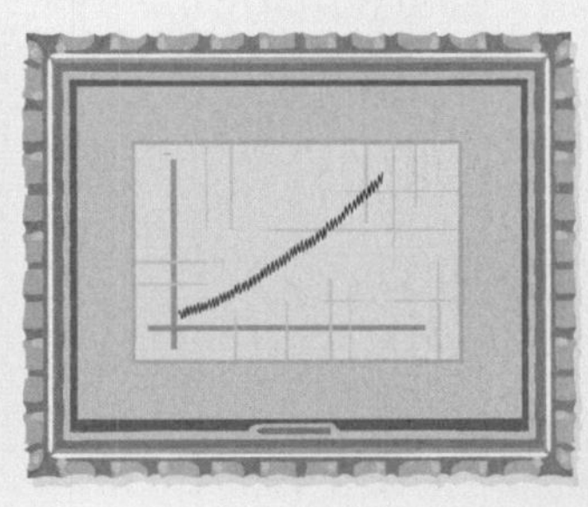

...AND ONE OF THE CENTRAL PIECES OF **EVIDENCE** IN THIS BOOK.

WHAT DOES IT **MEAN?**

IT MEANS **HUMAN ACTIVITY** IS VISIBLE ON A **PLANETARY SCALE.**

IN THE 19TH CENTURY THE MAJOR SOURCE OF HUMAN CO_2 EMISSIONS WAS **DEFORESTATION**...

CLEARING TREES TO MAKE ROOM FOR FARMS AND CITIES...
...RELEASES THE **CARBON** THAT WAS **STORED IN THOSE TREES** INTO THE ATMOSPHERE.

...BUT NOW IT'S THE **FOSSIL FUELS** THAT PROVIDE MOST OF OUR **ELECTRICITY** AND POWER OUR **FACTORIES** AND **CARS**.

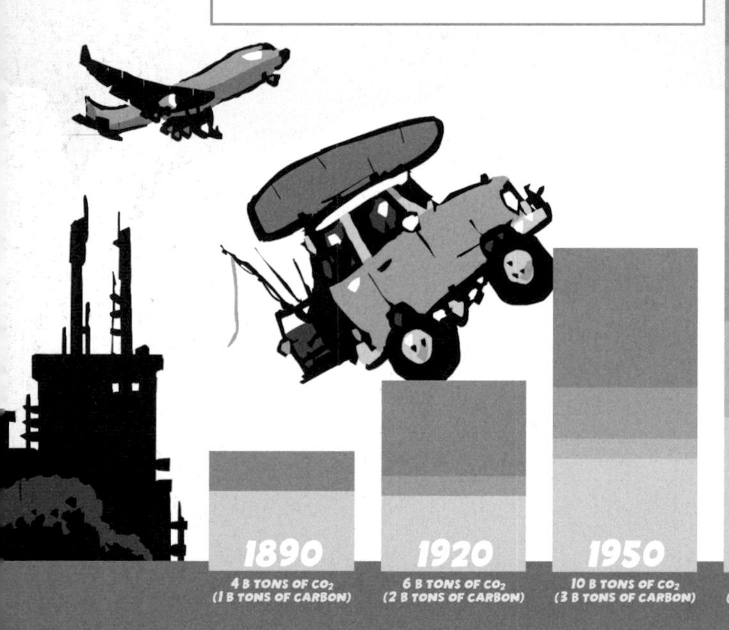

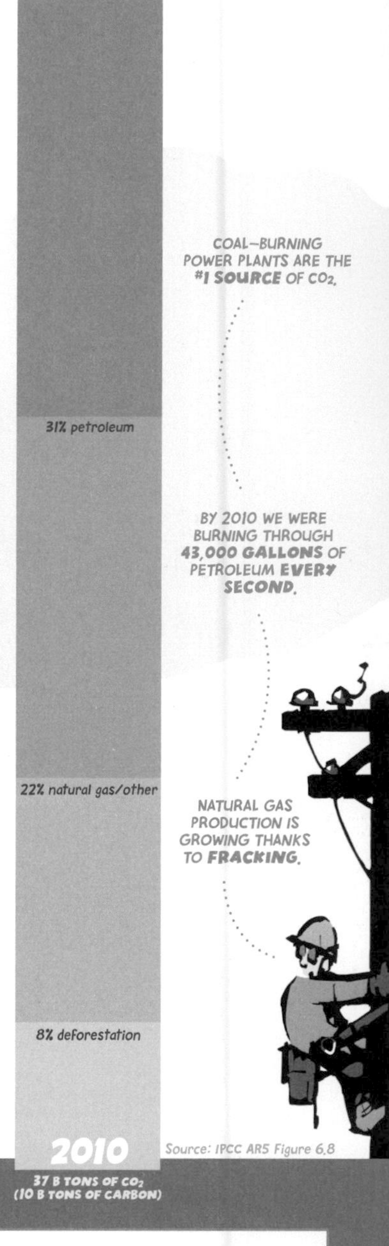
39% coal
31% petroleum
22% natural gas/other
8% deforestation
1890
1920
1950
1980
2010
4 B TONS OF CO_2 (1 B TONS OF CARBON)
6 B TONS OF CO_2 (2 B TONS OF CARBON)
10 B TONS OF CO_2 (3 B TONS OF CARBON)
24 B TONS OF CO_2 (7 B TONS OF CARBON)
37 B TONS OF CO_2 (10 B TONS OF CARBON)
Source: IPCC AR5 Figure 6.8
COAL-BURNING POWER PLANTS ARE THE **#1 SOURCE** OF CO_2.
BY 2010 WE WERE BURNING THROUGH **43,000 GALLONS** OF PETROLEUM **EVERY SECOND**.
NATURAL GAS PRODUCTION IS GROWING THANKS TO **FRACKING**.

ABOUT **60%** OF THAT EXTRA CO_2 GETS ABSORBED BY **PLANTS** AND **SOILS**...

...AND BY OTHER **CARBON SINKS** SUCH AS THE **OCEANS**...

FORESTS STILL COVER ABOUT **1/3** OF THE EARTH'S LAND.
WE'LL COME BACK TO THIS IN CHAPTER 15.

WE'LL COME BACK TO THIS IN CHAPTER 8, ON **OCEAN ACIDIFICATION**.

...BUT THE REST **STAYS IN THE ATMOSPHERE**, PUSHING UP THE KEELING CURVE.

HERE'S HOW THE **10 BILLION TONS** OF CARBON WE EMIT EACH YEAR AMPLIFIES THE **CARBON CYCLE.**

AS WE SAW ON PAGE 18, ABOUT **200 BILLION TONS** CYCLE NATURALLY EVERY YEAR.

BUT EVERY YEAR WE ADD ANOTHER **4 BILLION TONS** TO THE **ATMOSPHERE...**

...3 BILLION TONS TO THE **OCEANS...**

...AND 3 **BILLION TONS** THAT ARE ABSORBED BY **PLANTS AND SOILS.**

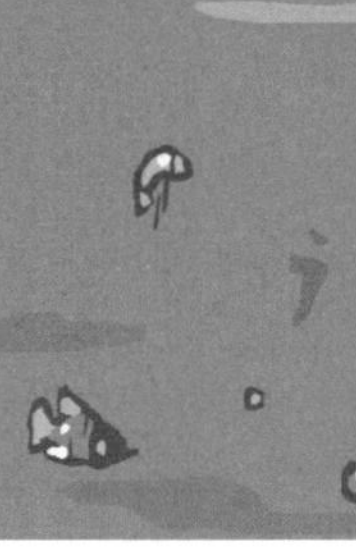

HUMAN EMISSIONS ARE A **SMALL FRACTION** OF THE NATURAL CYCLE, BUT OVER TIME **IT ADDS UP.**

INHALE MORE THAN YOU EXHALE...

...AND PRETTY SOON YOUR LUNGS WILL EXPLODE.

SCIENTISTS HAVE ALSO PUT THE KEELING CURVE IN **HISTORICAL CONTEXT**...

...BY STUDYING THE AMOUNT OF CO_2 TRAPPED IN **ICE CORES**.

WHAT DID CO2 CONCENTRATIONS LOOK LIKE **BEFORE 1958?**

DRILL, BABY, **DRILL**.

THEY DISCOVERED THAT **TODAY'S LEVELS** ARE **OFF THE CHARTS** COMPARED TO THE LAST 800,000 YEARS.

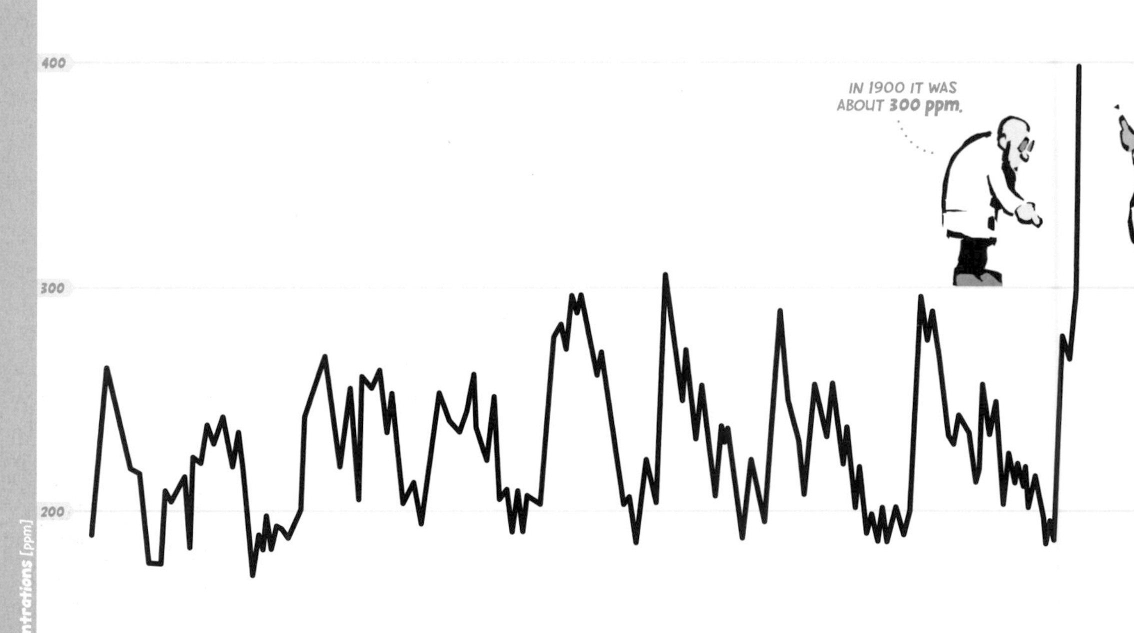
BY 2014 IT WAS NEAR **400 ppm**.
IN 1900 IT WAS ABOUT **300 ppm**.
400
300
200
CO2 concentrations [ppm]
800,000 ya
600,000 ya
400,000 ya
200,000 ya
Now

Source: IPCC AR5 Figure 5.3 and http://scrippsco2.ucsd.edu

EVEN **MORE AMAZING** IS THE RELATIONSHIP BETWEEN THIS **HISTORIC CO_2 DATA**...

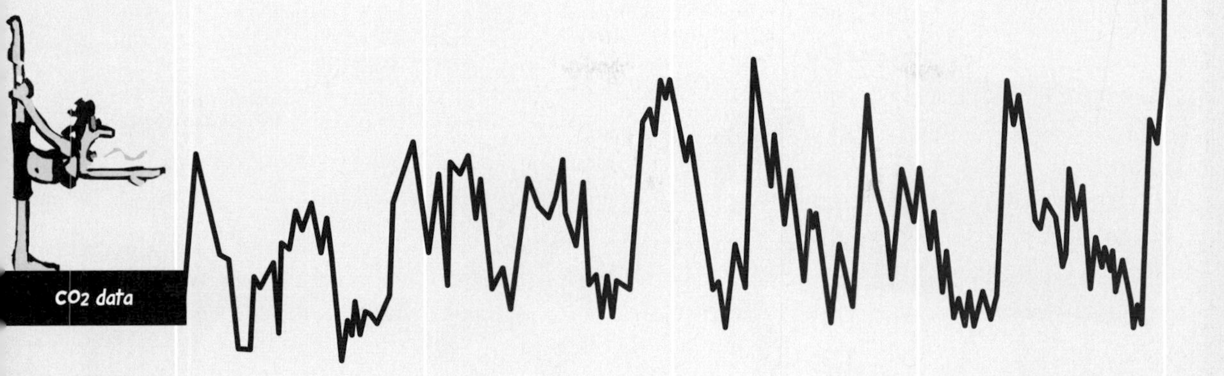

CO2 data

...AND THE **ICE CORE TEMPERATURE DATA** FROM PAGE 29.

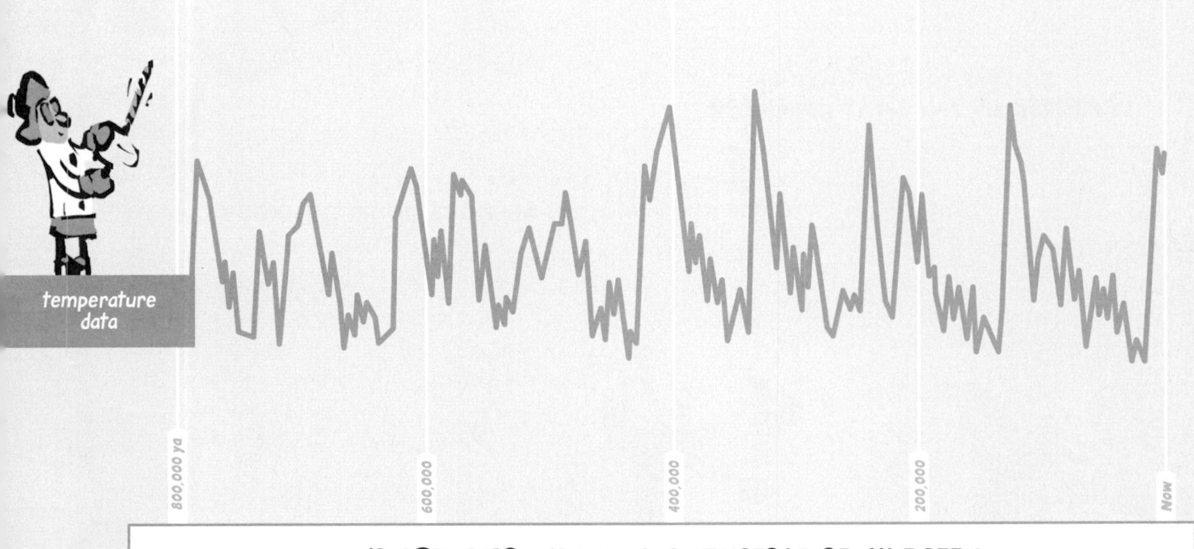

temperature data
800,000 ya
600,000
400,000
200,000
NOW

IT'S **OBVIOUS** THAT CO_2 AND TEMPERATURE HAVE BEEN **CLOSELY RELATED FOR EONS.**

WHAT ARE THE ODDS OF THAT HAPPENING BY CHANCE?
LESS THAN ONE IN A MILLION!

THE **CONNECTION** BETWEEN CO_2 AND TEMPERATURE IS SO STRONG THAT IT'S ***EASY TO GET CARRIED AWAY.***

DUDE, CO_2 MUST BE THE **DRIVING FORCE** BEHIND THE ICE AGES!
BE CAREFUL, THAT'S **NOT TRUE!**
CORRELATION IS **NOT CAUSATION.**

SO KEEP IN MIND WHAT WE **LEARNED IN THE LAST CHAPTER...**

...AS WE MOVE ON TO THE **NEXT CHAPTER.**

THE **MILANKOVITCH CYCLES SET THE TEMPO** FOR THE ICE AGES.

CO_2 AMPLIFIES THE MESSAGE!

Ice–Albedo Feedback
CO_2 Feedback

CHAPTER 5
ENERGY

YOUR BANK ACCOUNT DEPENDS ON MONEY IN AND MONEY OUT...
...AND THE EARTH'S TEMPERATURE DEPENDS ON ENERGY IN AND ENERGY OUT.

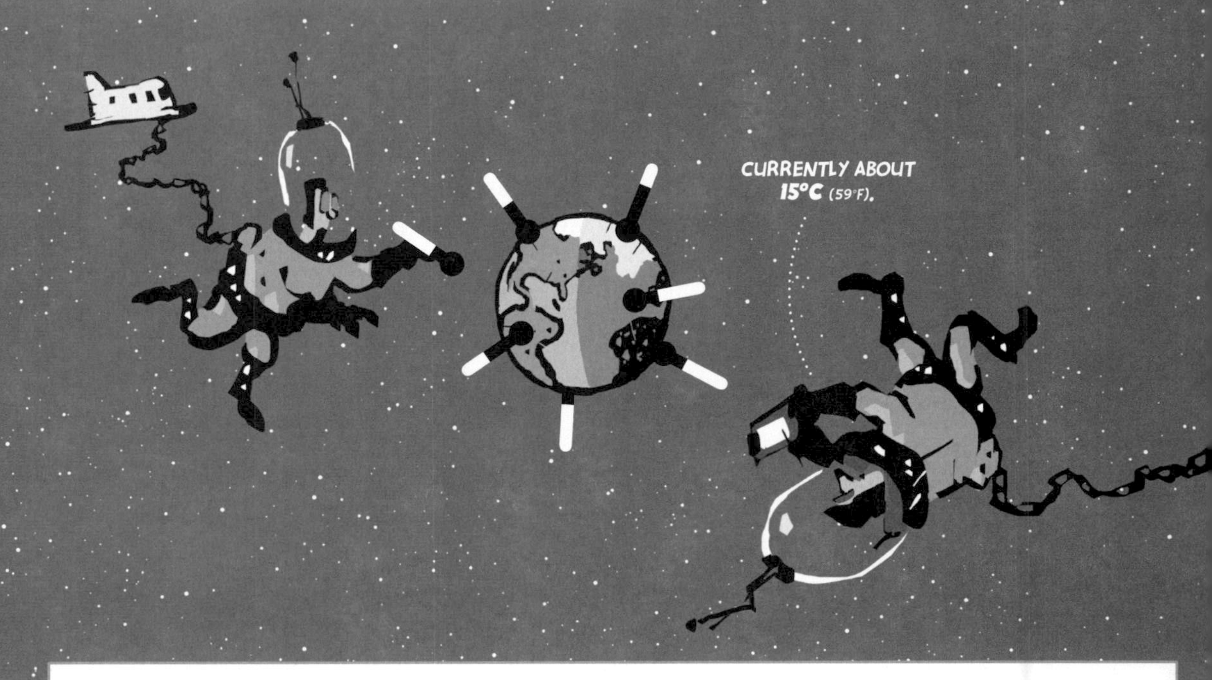
IF YOU SPEND A YEAR MEASURING AIR TEMPERATURES **ALL OVER THE SURFACE OF THE PLANET...**
CURRENTLY ABOUT **15°C** (59°F).
...YOU CAN CALCULATE THE **GLOBAL AVERAGE TEMPERATURE** FOR PLANET EARTH.

IF YOU WANTED TO, YOU COULD DO THE SAME THING FOR YOUR **HOUSE.**
WHAT ARE YOU **DOING?**
ABOUT **20°C** (68°F).

THE TWO DOMINANT INFLUENCES ON GLOBAL AVERAGE TEMPERATURE ARE **ENERGY IN** AND **ENERGY OUT**.

IT'S TRUE FOR YOUR HOUSE AND IT'S TRUE FOR THE WHOLE PLANET.
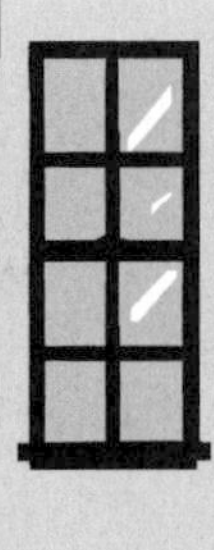

WHEN **ENERGY IN** EXCEEDS **ENERGY OUT**, THE PLANET **WARMS UP**...

JUST LIKE YOUR HOUSE WARMS UP WHEN YOU TURN ON THE HEATER.
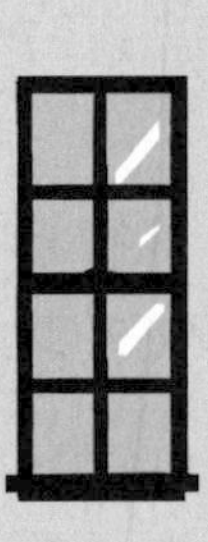

...AND WHEN **ENERGY OUT** EXCEEDS **ENERGY IN**, THE PLANET **COOLS DOWN**.

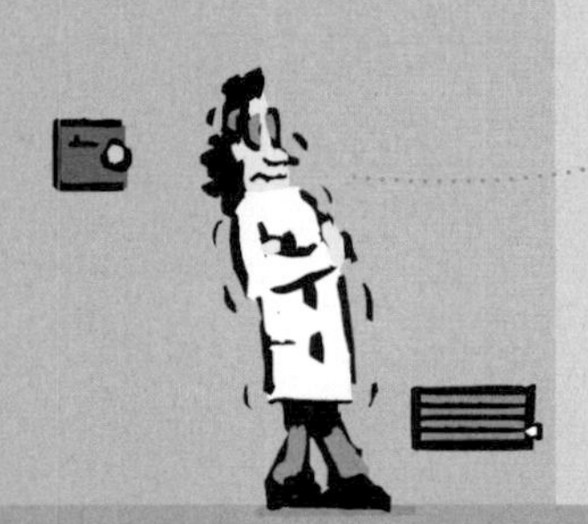
JUST LIKE YOUR HOUSE COOLS DOWN WHEN YOU OPEN A WINDOW.

SO TO UNDERSTAND GLOBAL TEMPERATURE WE NEED TO UNDERSTAND MORE ABOUT **ENERGY IN** AND **ENERGY OUT**.

ENERGY IN IS SIMPLE.

IF YOU PICTURE THE SUN AS A **BASKETBALL** ON ONE END OF A FULL-SIZE COURT...

...THEN THE EARTH WOULD BE A **LARGE GRAIN OF SAND** UNDER THE OTHER BASKET.

ENERGY FROM THE SUN COMES IN THE FORM OF **ELECTROMAGNETIC RADIATION**, WHICH INCLUDES...

...**ULTRAVIOLET (UV) RADIATION**, WITH SHORT, HIGH-ENERGY WAVELENGTHS...

UV IS PARTIALLY BLOCKED BY THE **OZONE LAYER**.

(SEE PAGE 20.)

...**VISIBLE LIGHT**, WHICH COVERS THE **WHOLE RAINBOW** (LITERALLY!)...

VISIBLE LIGHT IS THE ONLY KIND OF ELECTROMAGNETIC RADIATION PEOPLE CAN SEE.

...AND **INFRARED (IR) RADIATION**, WITH LONG, LOW-ENERGY WAVELENGTHS.

WE'LL COME BACK TO THIS IN A MOMENT.

ENERGY OUT IS MORE COMPLICATED.

FOR EXAMPLE, IT INCLUDES SOLAR ENERGY THAT IS **REFLECTED BACK INTO SPACE BY CLOUDS**...
...AND BY **THE EARTH'S SURFACE.**

IT'S ALBEDO!

ABOUT 30% OF ALL INCOMING ENERGY IS REFLECTED BY CLOUDS, ICE, SNOW, SAND, ETC.

CRUCIALLY, ENERGY OUT **ALSO** INCLUDES RADIATION **GIVEN OFF BY THE EARTH ITSELF.**

THE **EARTH** DOESN'T EMIT AS MUCH ENERGY AS THE **SUN**...
...BUT IT **DOES EMIT ENERGY.**

THIS OUTGOING ENERGY IS IN THE **INFRARED** PART OF THE ELECTROMAGNETIC SPECTRUM...

THE EARTH'S INFRARED WAVELENGTHS ARE EVEN **LONGER** AND **LOWER ENERGY** THAN INFRARED FROM THE SUN.

...AND IS FAMILIAR TO ANYBODY WHO HAS USED **THERMAL IMAGING** EQUIPMENT...

THERE'S A **LEAKY PIPE** INSIDE THIS WALL...

...AND A **MAN** HIDING IN THAT CLOSET.

...OR HUDDLED NEAR A **HOT FIRE.**

YOU CAN'T **SEE** INFRARED LIGHT, BUT YOU CAN **FEEL** IT.

IN THE MORNING IT WILL NO LONGER BE **RED HOT**...
...BUT IT WILL STILL BE **INFRARED HOT.**

AND THAT BRINGS US TO **GREENHOUSE GASES.**

WATER VAPOR
H2O

CARBON DIOXIDE
CO2

METHANE
CH4

Plus others that are less important.

WHAT MAKES THEM **GREENHOUSE** GASES IS THAT THEY **DON'T INTERACT** MUCH WITH **ENERGY IN**...

GO RIGHT ON IN.

DON'T MIND US.

CO2

...BUT **DO INTERACT** WITH **ENERGY OUT.**

WHOA!

SLOW DOWN.

H2O
CH4
CO2

GREENHOUSE GASES **BLOCK** SOME OF THE OUTGOING WAVELENGTHS.

BY REDUCING ENERGY OUT,
GREENHOUSE GASES **WARM THE PLANET.**

REDUCING ENERGY OUT IS ALSO HOW **INSULATION** WARMS UP YOUR **HOUSE**...
...OR A **BLANKET** WARMS UP YOUR **BODY**...
...OR A **GREENHOUSE** WARMS UP YOUR **TOMATOES.**

BY THE **1800s**, SCIENTISTS HAD FIGURED OUT THAT GREENHOUSE GASES HAVE A **MAJOR IMPACT** ON THE EARTH'S TEMPERATURE...
THE **ACTUAL** GLOBAL AVERAGE TEMPERATURE IS ABOUT **14°C** (57°F)...
... BUT **WITHOUT GREENHOUSE GASES** IT WOULD BE **−18°C** (0°F).

...AND TODAY, SCIENTISTS KNOW THAT SIMILAR GREENHOUSE EFFECTS EXIST **ON OTHER PLANETS**, LIKE **VENUS**.
CO_2
VENUS HAS **70 TIMES** MORE ATMOSPHERIC CO_2 THAN EARTH.
THAT'S A BIG REASON THE SURFACE TEMPERATURE IS **462°C** (863°F).
HOTTER THAN A **PIZZA OVEN!**

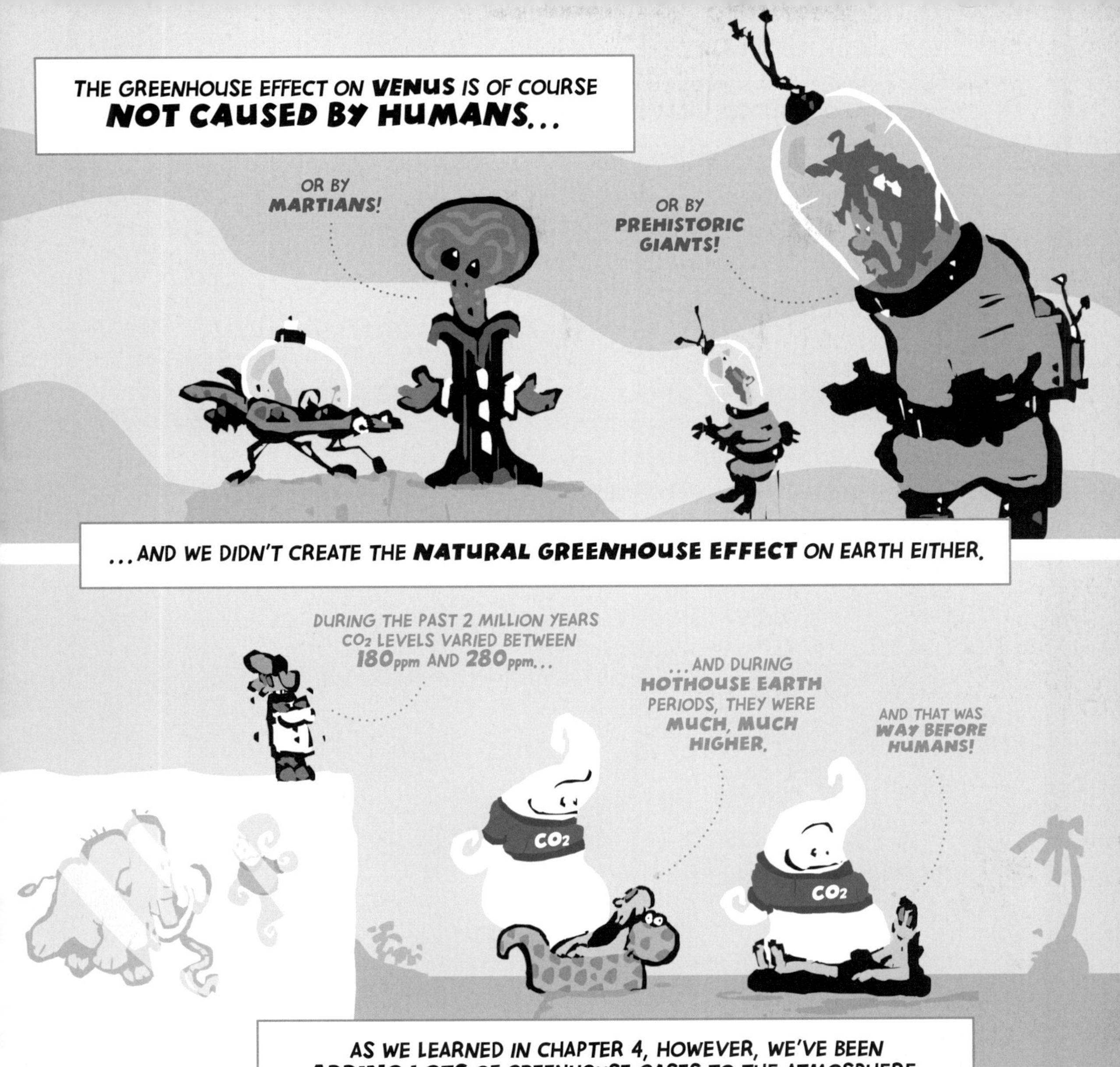
THE GREENHOUSE EFFECT ON VENUS IS OF COURSE NOT CAUSED BY HUMANS...
OR BY MARTIANS!
OR BY PREHISTORIC GIANTS!
...AND WE DIDN'T CREATE THE NATURAL GREENHOUSE EFFECT ON EARTH EITHER.
DURING THE PAST 2 MILLION YEARS CO_2 LEVELS VARIED BETWEEN 180ppm AND 280ppm...
...AND DURING HOTHOUSE EARTH PERIODS, THEY WERE MUCH, MUCH HIGHER.
AND THAT WAS WAY BEFORE HUMANS!
CO_2
CO_2

AS WE LEARNED IN CHAPTER 4, HOWEVER, WE'VE BEEN ADDING LOTS OF GREENHOUSE GASES TO THE ATMOSPHERE.
SINCE THE START OF THE INDUSTRIAL REVOLUTION IN THE LATE 1700s...
...WE'VE INCREASED CO_2 LEVELS FROM 280ppm...
...TO 400ppm.

THE SWEDISH CHEMIST **ARRHENIUS** WAS ONE OF THE FIRST PEOPLE TO SPECULATE ABOUT THE CONSEQUENCES OF AN **ENHANCED GREENHOUSE EFFECT.**

THIS IS GOING TO BE GREAT FOR FARMERS IN SWEDEN!

IN **1896** HE STUDIED WHAT WOULD HAPPEN IF WE **DOUBLED** CO_2 CONCENTRATIONS...

...AND MADE A ROUGH CALCULATION OF THE **EVENTUAL INCREASE** IN **GLOBAL AVERAGE TEMPERATURE.**

WHAT IF WE WENT FROM 280ppm...
...TO 560ppm?
ABOUT 5°C (9°F).

INCREDIBLY, HE CAME **VERY CLOSE** TO THE RANGE THAT CLIMATE SCIENTISTS CALCULATE TODAY.

ABOUT 1.5–4.5°C (2.7–8.1°F).

CHAPTER 6
CLIMATE SCIENCE

WHY SHOULD
I TRUST YOU?
DON'T TRUST US.
TRUST THE
SCIENTIFIC
METHOD.

THE **SCIENTIFIC METHOD** INVOLVES **DEVELOPING HYPOTHESES**...

FORCE EQUALS **MASS** TIMES **ACCELERATION**.
F=ma

...**TESTING** THEM **AGAINST THE REAL WORLD**...

YUP, IT WORKS ON **EARTH**.

AND ON THE **MOON**!

AND ON **MARS**!

...AND THEN **REFINING** THOSE HYPOTHESES TO INCORPORATE **NEW DATA**...

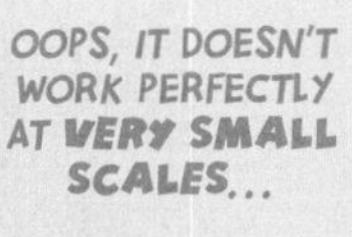
OOPS, IT DOESN'T WORK PERFECTLY AT **VERY SMALL SCALES**...

...OR AT **VERY HIGH SPEEDS**.

...AND **NEW IDEAS**.

LET'S TRY SOMETHING **DIFFERENT**.
E=mc
E=mc²
E=mc³

BECAUSE THIS PROCESS INVOLVES AN **ENDLESS SUPPLY OF NEW QUESTIONS**...

MOMMY, WHERE DO **ATOMS** COME FROM?
MOMMY, HOW OLD IS THE **UNIVERSE?**
MOMMY, WHY DO PRAYING MANTISES **EAT THEIR HUSBANDS?**

...AND **NEW EXPERIMENTS**...

LET'S SEE WHAT HAPPENS WHEN WE ADD **FROOT LOOPS!**

...IT WILL **NEVER** PRODUCE "**ABSOLUTE TRUTH.**"

EVEN THE **THEORY OF GRAVITY** IS **JUST A THEORY.**

BUT THE **SCIENTIFIC METHOD** IS STILL OUR **BEST TOOL** FOR **UNDERSTANDING THE WAY THE WORLD WORKS.**

IT'S WHAT YOU SHOULD **LOOK AT** BEFORE YOU **LEAP.**

THE IDEAL WAY TO **TEST A HYPOTHESIS** IS WITH A **CONTROLLED EXPERIMENT**...

WE GAVE **LOTS OF WATER** TO THESE PLANTS...
...AND **NO WATER** TO THESE ONES.
LET'S SEE **WHICH ONES SURVIVE.**

WE GAVE **CHILLAXITROL** TO THESE HYPERACTIVE KIDS...
...AND **NOT** TO THESE ONES.
LET'S SEE WHICH ONES CAN **CONCENTRATE BETTER.**

...BUT THAT'S **NOT ALWAYS POSSIBLE.**

FOR EXAMPLE, YOU CAN'T DO **CONTROLLED EXPERIMENTS** ABOUT **SMOKING...**

WE GAVE ADDICTIVE NICOTINE STICKS TO THESE KIDS...
...AND NOT TO THESE ONES.

...OR ABOUT **CLIMATE CHANGE.**

WE DOUBLED THE CO_2 ON THIS PLANET...
...AND NOT ON THIS ONE.

FORTUNATELY, THAT HASN'T STOPPED SCIENTISTS FROM **MAKING PROGRESS** ON BOTH OF THESE ISSUES.

WHEN YOU CAN'T WORK IN THE LAB...
...YOU HAVE TO GO OUTSIDE AND SEARCH FOR CLUES, LIKE A DETECTIVE.

IN FACT, THE SCIENTIFIC HISTORIES OF **SMOKING** AND **CLIMATE CHANGE** ARE SIMILAR.

STOP
WHAT **SIMILARITIES** COULD THERE POSSIBLY BE?

COUGH COUGH

OVER TIME, THE LINK BETWEEN **SMOKING** AND **DISEASE** HAS **GROWN CLEARER**...

"THE WEIGHT OF THE **EVIDENCE** SUGGESTS THAT EXCESSIVE SMOKING IS **ONE OF THE CAUSATIVE FACTORS** IN LUNG CANCER."

Surgeon General 1957

...AND **CLEARER**...

"CIGARETTE SMOKING IS **CAUSALLY RELATED** TO **LUNG CANCER** IN **MEN**."

Surgeon General 1964

...AND **CLEARER**...

"**CERVICAL** CANCER, **KIDNEY** CANCER, **PANCREATIC** CANCER..."

Surgeon General 2004

...AND SO HAS THE LINK BETWEEN **GREENHOUSE GAS EMISSIONS** AND **GLOBAL WARMING**.

"THE BALANCE OF **EVIDENCE** SUGGESTS A **DISCERNIBLE HUMAN INFLUENCE** ON GLOBAL CLIMATE."

IPCC 1995

"THERE IS **NEW** AND **STRONGER EVIDENCE**."

IPCC 2001

"IT IS **EXTREMELY LIKELY** THAT HUMAN INFLUENCE HAS BEEN THE **DOMINANT CAUSE** OF THE OBSERVED WARMING SINCE THE MID–20TH CENTURY."

IPCC 2013

THE **ULTIMATE TEST** OF A THEORY IS ITS ABILITY TO **PREDICT THE FUTURE**...

AS ANTICIPATED, SMOKING CAUSED OVER **400,000** DEATHS LAST YEAR IN THE UNITED STATES.

...AND SOME OF THE BEST EVIDENCE ABOUT **CLIMATE CHANGE** COMES FROM **SUCCESSFUL PREDICTIONS** FROM DECADES PAST.

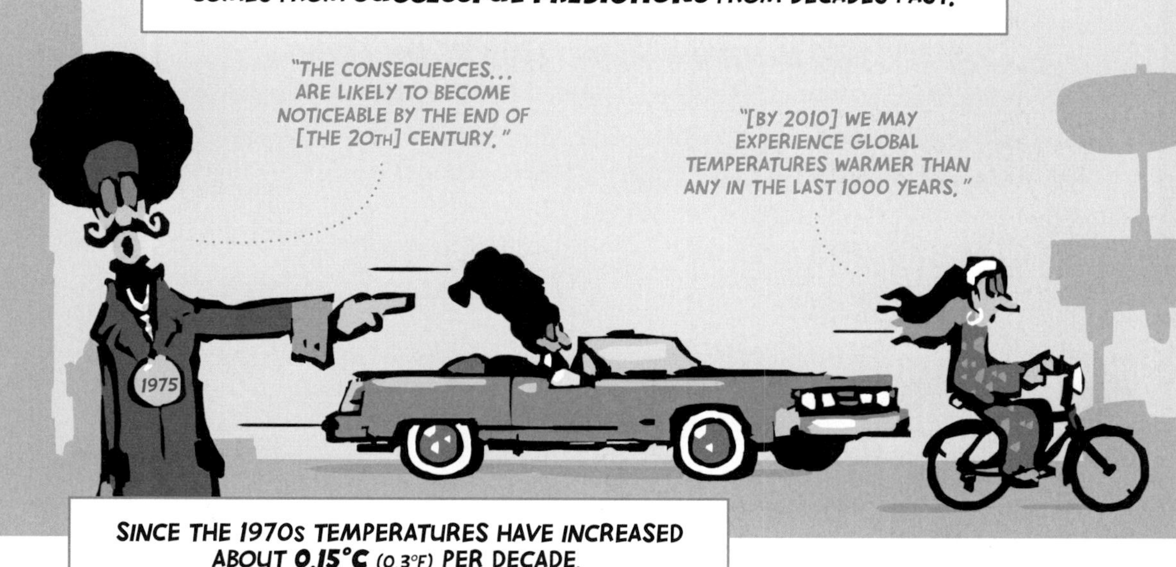

"THE CONSEQUENCES... ARE LIKELY TO BECOME NOTICEABLE BY THE END OF [THE 20TH] CENTURY."
1975
"[BY 2010] WE MAY EXPERIENCE GLOBAL TEMPERATURES WARMER THAN ANY IN THE LAST 1000 YEARS."

SINCE THE 1970S TEMPERATURES HAVE INCREASED ABOUT **0.15°C** (0.3°F) PER DECADE.

WE PREDICTED **0.2°C** (0.4°F).
NOT BAD!
OUR **FASHION SENSE** WAS **TERRIBLE**...
...BUT OUR PREDICTIONS WERE **GREAT**.
+0.6°C
+0.4°C
+0.2°C
1970
1980
1990
2000
2010
Temperatures relative to 20th century average.
Source: National Climatic Data Center, http://www.ncdc.noaa.gov/cag

AND CLIMATE SCIENTISTS HAVEN'T JUST BEEN **RIGHT ABOUT THE BIG PICTURE...**

... THEY'VE ALSO BEEN **RIGHT ABOUT LOTS OF THE DETAILS.**

MORE WARMING ON **LAND** THAN IN THE **OCEANS?**
YUP.

MORE WARMING **NEAR THE POLES** THAN **NEAR THE EQUATOR?**
YUP.

LESS ENERGY ESCAPING INTO **SPACE?**
YUP.

THOSE DETAILS ARE LIKE **FINGERPRINTS** AT A **CRIME SCENE.**

THIS HAS GOT **ANTHROPOGENIC GLOBAL WARMING** WRITTEN ALL OVER IT.

LIKE WE SAID, CLIMATE SCIENCE IS LIKE **DETECTIVE WORK...**

WHOEVER DID THIS MUST HAVE HAD THE **MEANS**, THE **MOTIVE**, AND THE **OPPORTUNITY**.

IT WASN'T ME, IT WAS **QUEEN VICTORIA**.

...AND **100 YEARS** OF **SCIENTIFIC DETECTIVE WORK...**

...HAS CONVINCED THE VAST MAJORITY OF SCIENTISTS THAT **HUMAN ACTIVITY IS THE DOMINANT CAUSE OF GLOBAL WARMING.**

WE'RE **CONVINCED**.
WE'RE **NOT**.
AAAS
IPCC
AMS
ACS
ASA
National Academies of Sciences
NIPCC

IPCC = INTERGOVERNMENTAL PANEL ON CLIMATE CHANGE; **AAAS** = AMERICAN ASSOCIATION FOR THE ADVANCEMENT OF SCIENCE; **AMS** = AMERICAN METEOROLOGICAL SOCIETY; **ACS** = AMERICAN CHEMICAL SOCIETY; **ASA** = AMERICAN STATISTICAL ASSOCIATION; **NATIONAL ACADEMIES** INCLUDES THE NATIONAL ACADEMIES OF THE U. S., BRAZIL, CANADA, CHINA, FRANCE, GERMANY, INDIA, ITALY, JAPAN, RUSSIA, THE U. K., AUSTRALIA, BELGIUM, THE CARIBBEAN, INDONESIA, IRELAND, MALASIA, NEW ZEALAND, AND SWEDEN; **NIPCC** = NONGOVERNMENTAL INTERNATIONAL PANEL ON CLIMATE CHANGE

OF COURSE, IT'S ALWAYS POSSIBLE THAT ALL THOSE SCIENTISTS ARE **WRONG.**

WHOOPS, GLOBAL WARMING IS ACTUALLY CAUSED BY BROCCOLI...
...AND SO IS LUNG CANCER...
...AND THE KILLER REALLY WAS QUEEN VICTORIA!

BUT SO FAR **NOBODY** HAS BEEN ABLE TO IDENTIFY A **GOOD COMPETING THEORY** FOR GLOBAL WARMING...

IS IT SUNSPOTS?
NOPE.

EL NIÑO?
NOPE.

MILANKOVITCH CYCLES?
NOPE.

BROCCOLI?
WHA?

...AND THE CASE FOR **ANTHROPOGENIC CLIMATE CHANGE** HAS GOTTEN **STRONGER AND STRONGER.**

IT'S ALL COMING TOGETHER.
THEORY
STATISTICAL ANALYSIS
REAL-WORLD EVIDENCE
SUCCESSFUL PREDICTIONS
ICE CORES
COMPUTER MODELS

MOST GLOBAL WARMING DATA IS PAINSTAKINGLY GATHERED **WITH FANCY EQUIPMENT...**

LIKE **SATELLITES...**

...OCEAN BUOYS...

...AND **ICE CORE DRILLS.**

...BUT IT'S POSSIBLE TO FIND EVIDENCE **ALL AROUND US.**

SNOW IS **MELTING EARLIER.**

FLOWERS ARE **BLOOMING EARLIER.**

BIRDS ARE **MIGRATING EARLIER.**

YOU MIGHT EVEN FIND THIS SORT OF **PHENOLOGICAL EVIDENCE** IN **YOUR OWN BACKYARD.**

A **PIED FLYCATCHER**, AND IT'S ONLY **APRIL 10!**

SOME **BIRDS** HAVE A NATURAL INSTINCT TO **MIGRATE...**
...AND SOME **HUMANS** HAVE A NATURAL INSTINCT TO **KEEP RECORDS.**

THESE DAYS, EVERYTHING WE KNOW ABOUT CLIMATE CHANGE IS INCORPORATED INTO CLIMATE MODELS RUN BY SUPERCOMPUTERS.
THESE MODELS DO A PRETTY GOOD JOB OF SIMULATING EVERYTHING FROM THE ICE AGES...
CHECK OUT CHAPTER 3.
...TO VOLCANIC ERUPTIONS...
SHORT-TERM COOLING FOR A FEW YEARS, BUT NO LONG-TERM IMPACT.
...AND THEY GIVE US THE BEST INDICATION OF WHAT TO EXPECT IN THE DECADES AHEAD.

PART TWO
PREDICTIONS

CHAPTER 7
GLOBAL WARMING

IS IT JUST ME...
...OR IS IT GETTING **HOT** IN HERE?!

NOW THAT WE'VE COVERED THE
BASIC SCIENCE OF CLIMATE CHANGE...
...LET'S TAKE A LOOK AT
WHAT MIGHT HAPPEN.
OF COURSE, LOTS OF THINGS COULD HAPPEN
IN THIS CENTURY AND BEYOND.

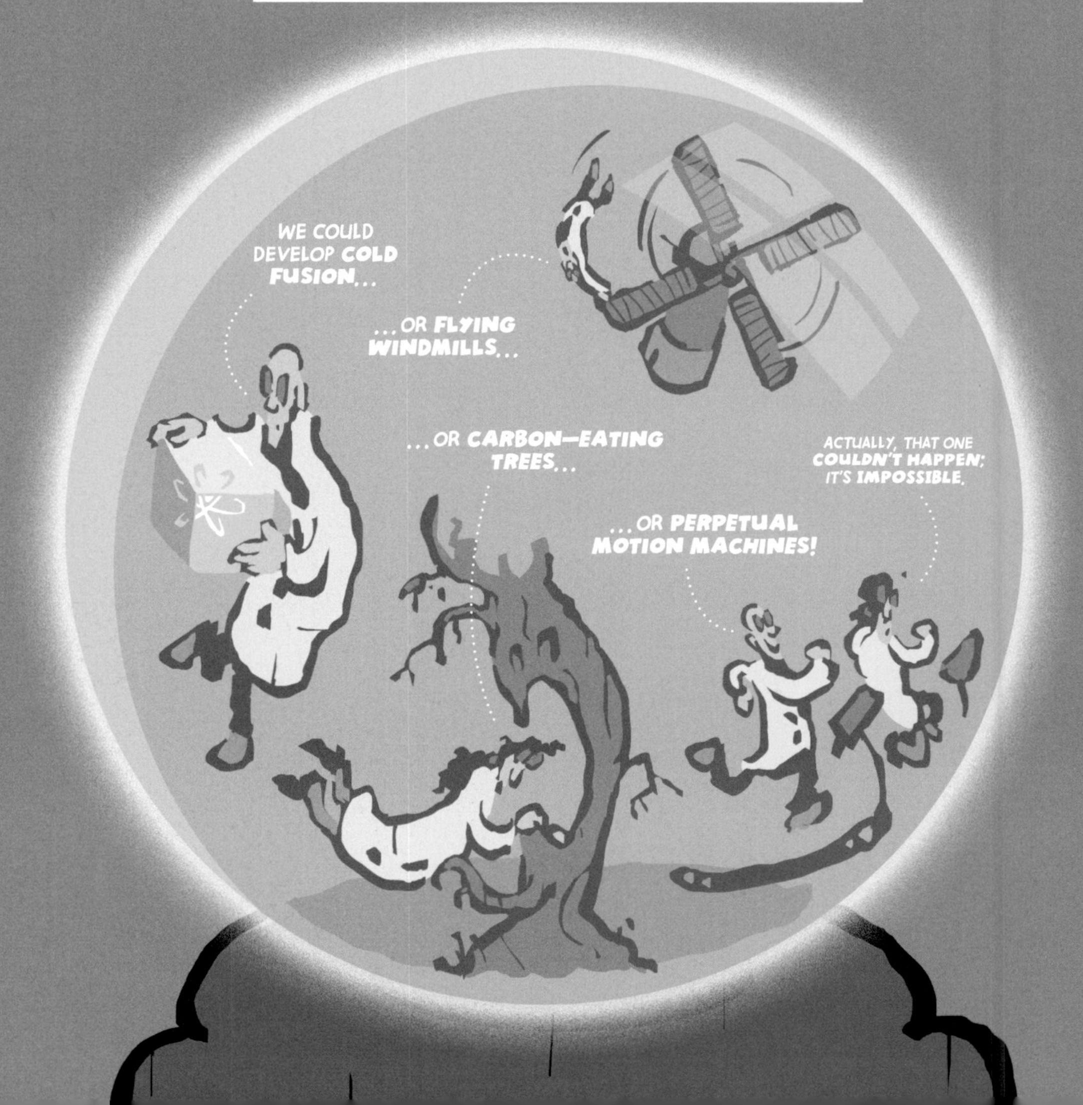
WE COULD
DEVELOP COLD
FUSION...
...OR FLYING
WINDMILLS...
...OR CARBON—EATING
TREES...
...OR PERPETUAL
MOTION MACHINES!
ACTUALLY, THAT ONE
COULDN'T HAPPEN;
IT'S IMPOSSIBLE.

ONE **VERY UNLIKELY** POSSIBILITY IS THAT HUMAN CARBON EMISSIONS WILL STOP **COLD TURKEY**.

THAT'S IT, **PARTY'S OVER**.
SHUT **EVERYTHING** DOWN **RIGHT NOW**.

A **MUCH MORE LIKELY** POSSIBILITY IS THAT **DEVELOPING COUNTRIES** LIKE CHINA AND INDIA WILL FOLLOW THE PATH BLAZED BY **DEVELOPED COUNTRIES** LIKE THE U.S.

FOSSIL FUELS TO POWER **HOMES, FACTORIES**, AND **MOTOR VEHICLES**...
...AND **DEFORESTATION** TO CLEAR LAND FOR FARMS, ROADS, AND BUILDINGS.

WE CALL THIS PATH **BUSINESS AS USUAL**.

WHAT IF WE **DIDN'T DO MUCH** TO REDUCE CARBON EMISSIONS?
THAT WOULD BE **BUSINESS AS USUAL**.

THE KEY FACT ABOUT **BUSINESS AS USUAL** IS THAT IT WOULD MAKE CO_2 EMISSIONS

EXPLODE.

TO SEE WHY, NOTE THAT WE CAN NEATLY DIVIDE THE WORLD'S 7 BILLION PEOPLE INTO **5 CHUNKS THE SIZE OF CHINA.**

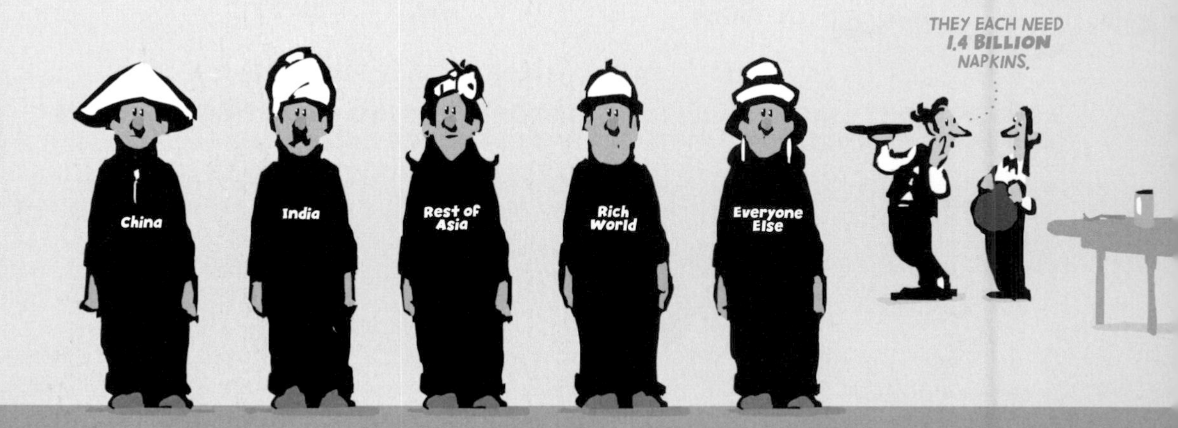
China
India
Rest of Asia
Rich World
Everyone Else
THEY EACH NEED 1.4 BILLION NAPKINS.

THE RICH WORLD MAKES UP **JUST ONE** OF THOSE "FIVE CHINAS"...

...BUT AT THE START OF THE 21ST CENTURY RICH WORLD ECONOMIES WERE RESPONSIBLE FOR ABOUT **HALF OF THE WORLD'S FOSSIL FUEL CONSUMPTION.**

ONE CAKE FOR YOU.
YOU ALL CAN SHARE THE SECOND CAKE.

UNDER **BUSINESS AS USUAL**...

THANKS TO RAPID ECONOMIC GROWTH...
...WE CAN NOW AFFORD MORE CAKE!

...THE OTHER FOUR CHINAS COULD **CATCH UP** BY **2100**.

PLUS, POPULATION GROWTH COULD **ADD ANOTHER 2 CHINAS TO THE PLANET.**

THAT'S THREE MORE CAKES.
THAT'S ANOTHER TWO CAKES.
Africa
More Asia

THAT MEANS THAT OVER THE COURSE OF THE CENTURY WE'D **INCREASE OUR FOSSIL FUEL CONSUMPTION BY 250%.**

THAT'S LIKE GOING FROM 2 CAKES A YEAR...

...TO 7.

SURE ENOUGH, UNDER BUSINESS AS USUAL, EMISSIONS COULD RISE **250%** DURING THIS CENTURY...
1900
ABOUT 4 B TONS OF CO_2 (ABOUT 1 B TONS OF CARBON)
1950
10 B TONS OF CO_2 (3 B TONS OF CARBON)
2000
30 B TONS OF CO_2 (8 B TONS OF CARBON)
2050
75 B TONS OF CO_2 (20 B TONS OF CARBON)
2100
105 B TONS OF CO_2 (28 B TONS OF CARBON)
900
800
700
600
500

Source: IPCC AR5 Figures TS.19 (above) and 5.3 (below)

...PUSHING ATMOSPHERIC CO_2 CONCENTRATIONS UP NEAR **1000 ppm.**

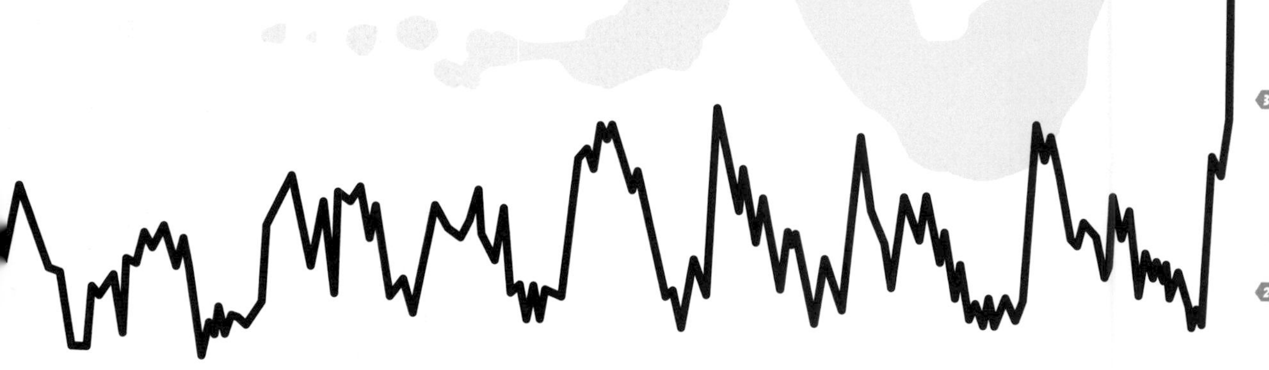
400
300
200

CO_2 CONCENTRATIONS HAVEN'T BEEN THAT HIGH FOR ***MILLIONS OF YEARS.***
100
600,000 ya
400,000 ya
200,000 ya
Now

AS A RESULT, THE GLOBAL AVERAGE TEMPERATURE CHANGE BETWEEN 2000 AND 2100...
...IF WE DON'T DO MUCH TO REDUCE EMISSIONS...
...WILL LIKELY BE ABOUT 4°C (7°F).
THAT ESTIMATE COULD BE A BIT HIGH...
MAYBE IT'LL BE ONLY 3°C.
...OR MORE THAN A BIT LOW...
MAYBE AS MUCH AS 5° OR 6°C.
MEH.
WE'LL COME BACK TO UNCERTAINTY IN CHAPTER 11.
...BUT IT MAKES A GOOD STARTING POINT FOR UNDERSTANDING THE CHANGES THAT ARE COMING.

IT'S JUST A NUMBER...
...WHAT EXACTLY DOES IT MEAN?

FOR SOME PEOPLE, A GLOBAL INCREASE OF 4°C SOUNDS LIKE **NO BIG DEAL**...

I'LL JUST PAY A BIT MORE FOR AIR CONDITIONING IN THE SUMMER...
...AND A BIT LESS FOR HEATING IN THE WINTER.
WHY ALL THE FUSS?

...WHILE TO OTHERS IT SOUNDS LIKE **A BIG, BIG DEAL.**

IN MOST PLACES ON EARTH, THE AVERAGE SUMMER IN 2100...
...WILL BE HOTTER THAN THE HOTTEST SUMMER OF THE 20TH CENTURY!

BUT NEITHER OF THESE IS THE **BEST WAY TO THINK ABOUT IT.**

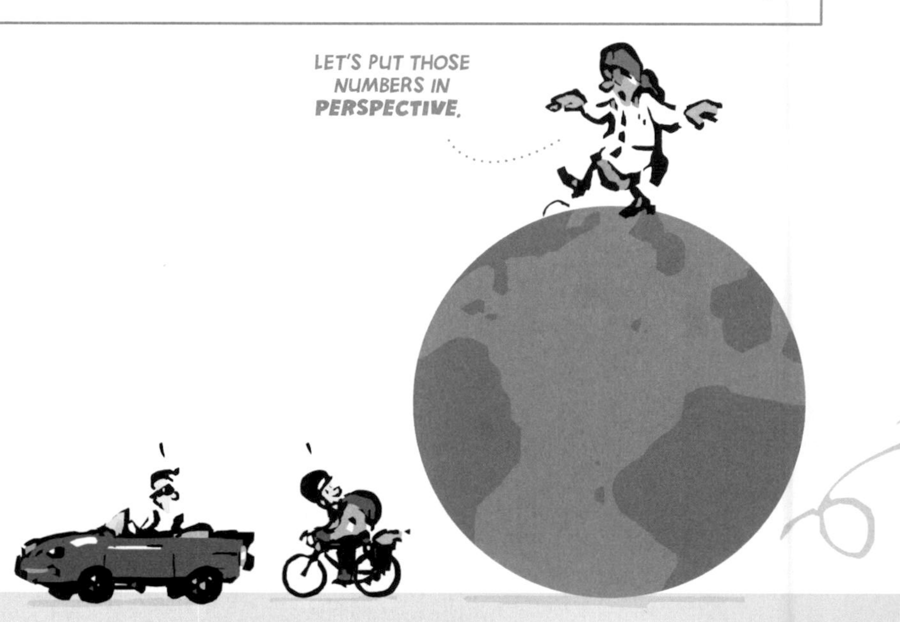
LET'S PUT THOSE NUMBERS IN PERSPECTIVE.

THE **BEST** WAY TO THINK ABOUT **4°C** IS LIKE
THE TITLE OF A HUGE BOOK.

4°C
A User's Guide to the Consequences of Business as Usual Between Now & 2100
AND BE CAREFUL ABOUT JUDGING A BOOK BY ITS COVER.
5°C
6°C
350 ppm
600 ppm
1000 ppm
RCP 2.6
RCP 4.5
RCP 6.0
RCP 8.5
Cold Turkey
Warm Turkey
Goose Egg
Chicken Little
Fowl Weather
Similar Titles in Climate Science

INSIDE THE BOOK ARE DESCRIPTIONS OF
LOTS OF DIFFERENT CHANGES...

...IN **LOTS OF DIFFERENT PLACES.**

TELL ME ABOUT SEA LEVEL RISE IN ASIA.
ONE CHAPTER A NIGHT AND WE'LL BE DONE BY THE TIME YOU GO TO COLLEGE.
4°C

FOR EXAMPLE, A GLOBAL INCREASE OF **4°C** (7°F) WOULD LIKELY WARM **OCEANS** BY ONLY **3°C** (5°F)...

...**LAND AREAS** BY MORE, AROUND **5°C** (9°F)...

OUCH OUCH OUCH OUCH.

...AND THE **ARCTIC** BY A WHOPPING **8°C** (15°F).

LESS ICE MEANS MORE ABSORPTION OF SOLAR ENERGY.

IT'S THE ICE–ALBEDO EFFECT!

AND THAT'S JUST THE **TIP OF THE ICEBERG.**

Welcome to the
ARCTIC CIRCLE
Titanic 2.0
"Well, there used to be icebergs here."

THE FACT IS, A GLOBAL INCREASE OF **4°C** (7°F) WOULD **TRANSFORM** THE **ENTIRE WORLD**...

Auf Wiedersehen
FROSTBITE

Hallo
HEATSTROKE

Salaam from the FERTILE FARMLAND of
BANGLADESH
"Well, there used to be farmland here."

WHEN IT COMES TO
LATITUDE:
50
IS THE NEW
40

SMOKEY THE BEAR KNOWS...
...YOU GET MORE FOREST FIRES
WHEN IT'S HOT AND DRY.

Monsoons just got...
Monsoonier

The World's Deserts
NOW EVEN BIGGER & DRIER!

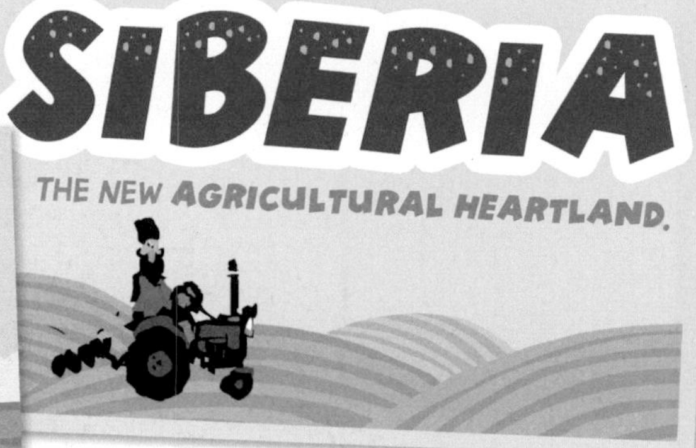
SIBERIA
THE NEW AGRICULTURAL HEARTLAND.

Swim the
Great
Barrier
Reef
HURRY, Offer ENDS SOON!

...BECAUSE WHEN THE CLIMATE CHANGES, ***EVERYTHING CHANGES.***

WE'RE GOING TO SPEND THE NEXT TWO CHAPTERS LEARNING ABOUT SOME OF THESE CHANGES IN **GREATER DETAIL.**

WE'LL START WITH THE CHANGES THAT HAVE TO DO WITH WATER.
Critics Rave about 4°C
"A Hot Topic..."
"Steamy and Stormy."
"A Wild Ride!"
"Insane!"

CHAPTER 8
WATER

WATER COVERS **MOST** OF THE EARTH'S SURFACE...

...AND THE **WATER CYCLE**...

WATER CONTINUOUSLY **EVAPORATES** INTO THE ATMOSPHERE...

...WHERE IT LINGERS AS CLOUDS AND WATER VAPOR BEFORE **PRECIPITATING** AS RAIN AND SNOW.

SOME WATER ACCUMULATES ON LAND AS **SNOW** AND **ICE** AND IN **LAKES**...

...AND SOME SEEPS **UNDERGROUND**...

...BUT **MOST** RETURNS TO **THE OCEANS**.

97% OF THE EARTH'S WATER IS STORED IN THE OCEANS.

...IS AS CRUCIAL TO **LIFE ON EARTH** AS THE CARBON CYCLE.

Mostly Water and Carbon

Mostly Water and Carbon

YOU **ARE** WHAT YOU **DRINK!**

Mostly Water and Carbon

WATER AND **CLIMATE** ARE ALSO CLOSELY RELATED.

HUMID AIR AND **OCEAN CURRENTS** MOVE HEAT FROM THE EQUATOR TOWARD THE POLES.
LIKE A **COOLING** AND **HEATING** SYSTEM!

SO IT'S NOT SURPRISING THAT MANY OF THE **MOST IMPORTANT IMPACTS OF CLIMATE CHANGE...**

SEA LEVEL RISE

FLOODS AND DROUGHTS

OCEAN ACIDIFICATION

...ARE RELATED TO THE **THREE FORMS OF WATER.**

LIQUID
WATER

SOLID
ICE AND SNOW

GAS
WATER VAPOR

LIQUID
WATER

WE BEGIN WITH THE **OCEANS**, WHICH ARE **WARMING UP** JUST LIKE THE REST OF THE PLANET.

OVER 90% OF THE EXTRA ENERGY TRAPPED BY GREENHOUSE GASES **ENDS UP IN THE OCEANS.**

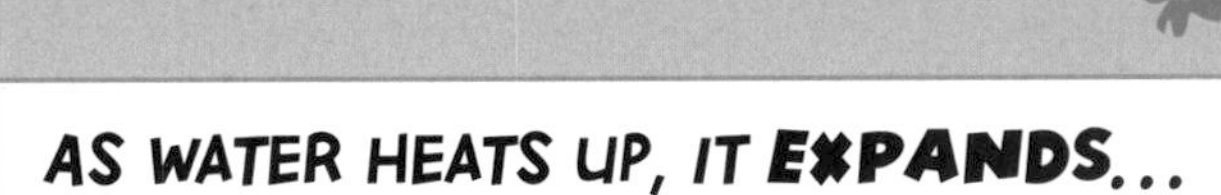

AS WATER HEATS UP, IT **EXPANDS...**

LIKE **MERCURY** IN A **THERMOMETER...**

...OR HOT AIR IN A **BALLOON.**

...AND THE RESULT IS **RISING SEA LEVELS.**

THERMAL EXPANSION ACCOUNTED FOR ABOUT **HALF** OF THE 7 INCHES OF SEA LEVEL RISE WE SAW DURING THE 20TH CENTURY.

THE OTHER HALF WAS FROM **MELTING GLACIERS AND ICE SHEETS.**

SEA LEVEL RISE IS LIKELY TO **ACCELERATE** IN THE YEARS AHEAD, POSING RISKS TO **AGRICULTURE**...

INSTEAD OF WORRYING ABOUT **WEEDS**...
...I'M WORRIED ABOUT **SEAWEEDS**.

...TO **LOW-LYING CITIES**...

...AND EVEN TO **ENTIRE COUNTRIES**.

Welcome to the

Maldive Islands

"Well, there **used to be islands** here."

ALL TOLD, **BUSINESS AS USUAL** WILL LIKELY LEAD TO ABOUT **2 FEET OF SEA LEVEL RISE THIS CENTURY**.

BUT IT COULD BE **TWO OR THREE TIMES** AS MUCH...
...DEPENDING ON WHAT HAPPENS TO **ICE AND SNOW**.

ICE AND SNOW

THE **ICE SHEETS** OF **GREENLAND** AND **ANTARCTICA** ARE CRUCIAL TO SEA LEVEL RISE THIS CENTURY.

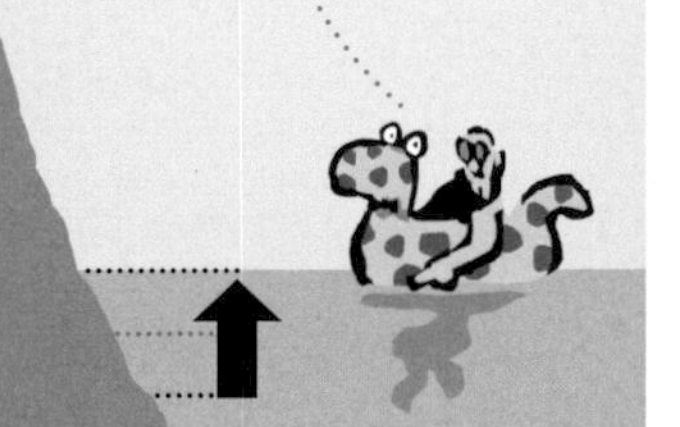
IF THEY MELT **SLOWLY**, SEA LEVEL RISE WILL BE ABOUT **2 FEET**.

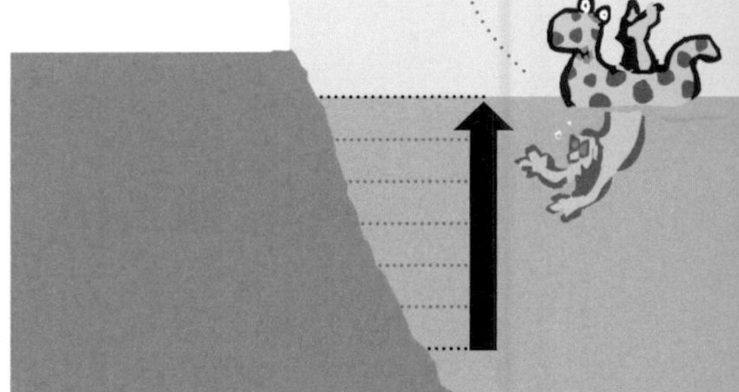
IF THEY MELT **QUICKLY**, IT COULD BE AS MUCH AS **6 FEET**.

IN CONTRAST, **MELTING ICE IN THE ARCTIC** DOES **NOT** RAISE SEA LEVELS BECAUSE IT'S **FLOATING**...

WHEN FLOATING ICE MELTS...
...THE **WATER LEVEL DOESN'T CHANGE**.

YOU CAN WATCH IT HAPPEN IN YOUR **DRINK**.

...BUT IT DOES MEAN **BIG CHANGES IN THE ARCTIC ENVIRONMENT**.

Welcome Back to the
ARCTIC CIRCLE
"An emerging Epicenter of Industry and Trade akin to the Mediterranean Sea."

GLOBAL WARMING ALSO MEANS A **SHIFT FROM SNOW TO RAIN**...

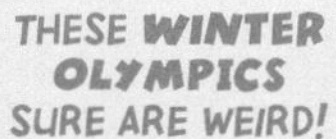
THESE **WINTER OLYMPICS** SURE ARE WEIRD!

MAYBE WE SHOULD TRY **WATER SKIING** INSTEAD.

...AND **EARLIER MELTING** OF THE SNOW THAT DOES FALL.

OF COURSE, **SNOW ISN'T JUST FOR SKIING**...

IT'S ALSO A **NATURAL RESERVOIR** THAT STORES WATER FROM THE **WINTER**...

...AND RELEASES IT IN THE **SUMMER**.

...SO LACK OF SNOW COULD MEAN BIG TROUBLE FOR **FARMERS, FAMILIES, AND INDUSTRY**.

ESPECIALLY WHEN COMBINED WITH **CHANGES IN RAINFALL**...
...WHICH IS OUR NEXT TOPIC.

WATER VAPOR
WE LEARNED IN CHAPTER 5 THAT WATER VAPOR IN THE ATMOSPHERE WORKS AS A GREENHOUSE GAS TO WARM THE PLANET.
ENERGY IN?
GO RIGHT AHEAD.
H2O
ENERGY OUT?
YOU SHALL NOT PASS!
H2O

BUT IT TURNS OUT THAT AS THE PLANET WARMS, THE ATMOSPHERE HOLDS MORE WATER VAPOR.
THE CLAUSIUS–CLAPEYRON RELATIONSHIP...
...SAYS THAT EACH ADDITIONAL 1°C MEANS 7% MORE WATER VAPOR.
HEAT AND HUMIDITY GO HAND IN HAND.
AND THAT CREATES YET ANOTHER POSITIVE FEEDBACK LOOP.

HOT AND HUMID TURNS INTO EVEN MORE HOT AND HUMID.

MORE HUMIDITY MAKES IT HOTTER...
MORE HEAT MAKES IT MORE HUMID...

GLOBAL WARMING **ALSO MEANS** BIG CHANGES IN **PRECIPITATION.**

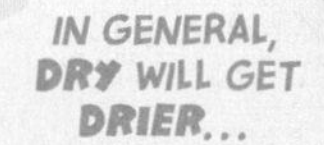

IN GENERAL, DRY WILL GET DRIER...

...AND WET WILL GET WETTER.

AS THE PLANET WARMS, THERE WILL BE **MORE EVAPORATION FROM THE SURFACE...**

...WHICH MEANS MANY **DRY** PLACES WILL GET **DRIER.**

I'M SWEATY.

I'M THIRSTY.

BUT THE EXTRA MOISTURE IN THE ATMOSPHERE **WILL ALSO CAUSE MORE INTENSE RAINSTORMS...**

...SO MANY **WET** PLACES WILL GET **WETTER.**

THE WATER CYCLE IS GOING CRAZY!

AN ADDITIONAL 4°C MEANS ABOUT 28% MORE WATER VAPOR IN THE ATMOSPHERE.

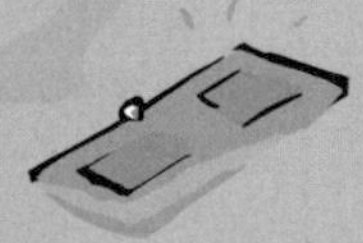

BONUS
LIQUID WATER

FINALLY, WE RETURN TO **THE OCEANS.**

THEY'RE NOT JUST **WARMING UP** AND **EXPANDING**...
...THEY'RE ALSO GETTING **MORE ACIDIC.**

OCEAN ACIDIFICATION HAPPENS WHEN CO_2 FROM THE ATMOSPHERE **DISSOLVES IN OCEAN WATER**...

RECALL FROM PAGE 47 THAT OVER **25%** OF OUR CO_2 EMISSIONS END UP **DOWN THERE.**
AND RECALL THIS FROM **CHEMISTRY CLASS!**
$CO_2 + H_2O \rightarrow H_2CO_3 \rightarrow H^+ + HCO_3^-$
SHOW OFF!

...INCREASING THE CONCENTRATION OF **HYDROGEN IONS** IN THE WATER.

MORE HYDROGEN IONS (H^+)...
...MEANS **MORE ACIDIC.**

Bleach
pH=12.6

Sea Water
pH=8

Coca-Cola
pH=2.52

INCREASED ACIDITY ERODES **CORAL REEFS**...
EXTRA HYDROGEN IONS BREAK DOWN THE REEF **FASTER THAN THE CORAL CAN BUILD IT UP.**
...AND **DISSOLVES** THE **SKELETONS** AND **SHELLS** OF SEA CREATURES, BOTH **BIG**...

YUCK.

DOUBLE YUCK.

...AND **SMALL.**

COCCOLITHOPHORES

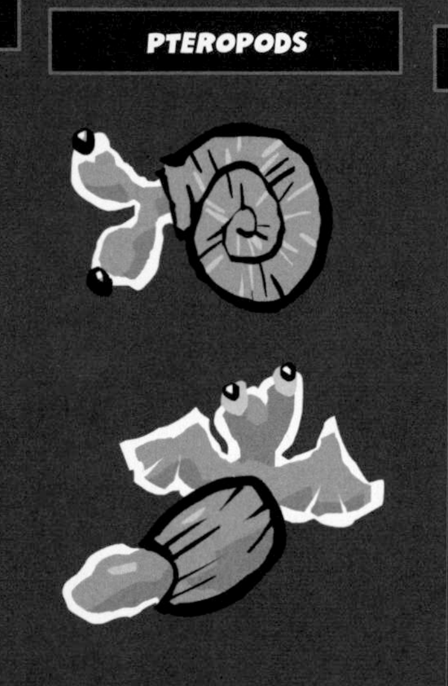
PTEROPODS

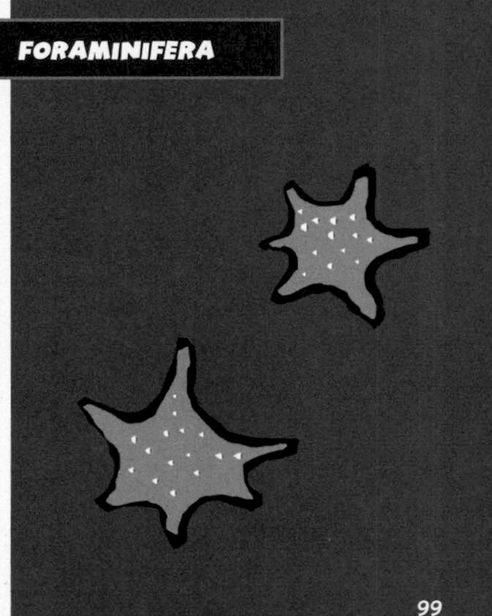
FORAMINIFERA

EVEN THOUGH YOU'VE **PROBABLY NEVER HEARD OF THEM...**

CACALITHOFARTS!
FORBADIFADIDALIDA!
PHPHPHTHTHTHBBBB!

... THOSE TINY SEA CREATURES FORM THE **BASE OF THE MARINE FOOD CHAIN.**

AS A RESULT, OCEAN ACIDIFICATION THREATENS JUST ABOUT **EVERYTHING** IN THE SEA...

... AND **EVERYONE** CONNECTED TO IT.

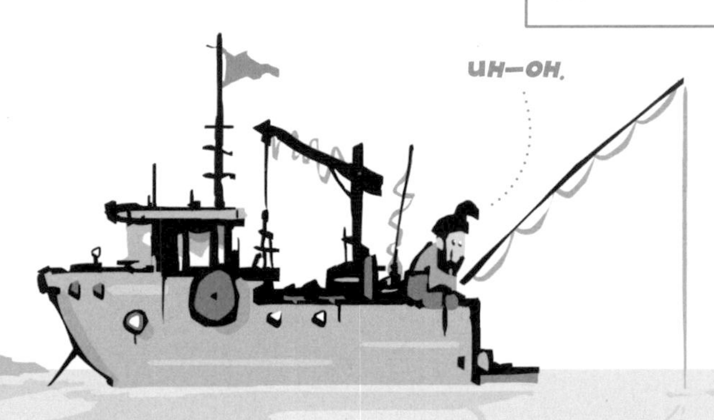
UH—OH.

IT'S NO WONDER **OCEAN ACIDIFICATION** AND **GLOBAL WARMING** ARE SOMETIMES CALLED **EVIL TWINS.**

LET'S PUT COCA—COLA IN THE BUCKET AND WATCH THE SHELLFISH DISSOLVE.
WE'RE DOUBLE TROUBLE!

CHAPTER 9
LIFE ON EARTH

WHEN CIRCUMSTANCES CHANGE, LIVING THINGS **ADAPT**...

SALE!
SORRY, NO **SEAL MEAT** TODAY.

IN THAT CASE, GIVE ME 50 POUNDS OF THE **SNOW GOOSE**.

...AND SPECIES **EVOLVE**...

GO BACK **100,000 GENERATIONS** AND YOUR ANCESTORS WERE JUST REGULAR OLD **BROWN BEARS**.

...AND IF THEY DON'T SUCCEED **IT'S CURTAINS.**

WHAT USE ARE BIG WEBBED FEET AND TRANSPARENT FUR IF THERE'S **NO MORE ICE?**
THAT'S WHY YOU NEED TO **GO TO SCHOOL** AND **PAY ATTENTION!**

MANY OF THE CHANGING CIRCUMSTANCES IN THE WORLD TODAY ARE CAUSED BY HUMANS...
BOY IS THAT AN UNDERSTATEMENT!
HABITAT LOSS
....AND CLIMATE CHANGE IS GOING TO ADD AN EXTRA CHALLENGE.
INVASIVE SPECIES
POLLUTION
POPULATION INCREASE
OVERFISHING

CLIMATE CHANGE IS LIKELY TO BE A BIG PROBLEM FOR SPECIES THAT **EVOLVE SLOWLY**...

WE TAKE **EVERYTHING** SLOWLY.

...AND SCIENTISTS ESTIMATE THAT **40–70%** OF ALL THE SPECIES ON EARTH MAY GO **EXTINCT** OVER THE NEXT 100 YEARS.

Armored slowpoker

Invasivus geurrillavinus

Acreus needsus

Annoying nevergoextinctus

Wartius allergictofungi

Scavengus canliveanywhereus

Albino nonvegetarianix

Giganticus pickyeater

Omnivorous gourmand

Rex thumperhumper
Execrable philofeceus

Spectacularus useless

Gracious vulnerabilus

Humanus engineerdus

A CHANGING WORLD MAY BE LESS OF A PROBLEM FOR SPECIES THAT **EVOLVE QUICKLY**...

WE BREED LIKE **RABBITS!**

...OR CAN **ADAPT QUICKLY**...

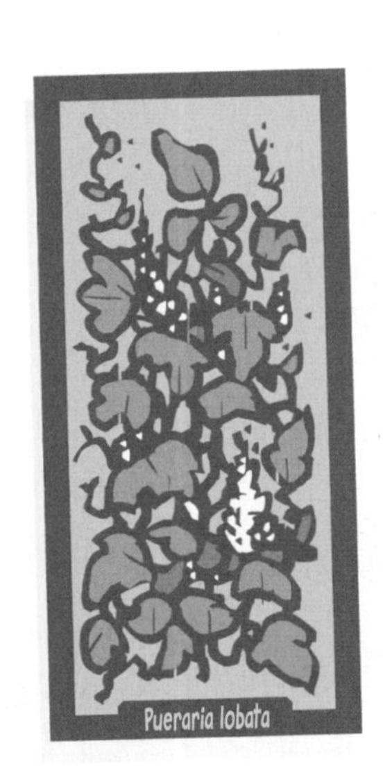
Pueraria lobata

Rattus norvegicus

Columba livia

Homo sapiens

HEY!

...OR MIGHT **BENEFIT** FROM THE NEW REALITY.

CERTAIN PLANTS DO **BETTER** WITH MORE CO_2...
...AS LONG AS IT DOESN'T GET **TOO HOT**.

AS FOR HOW CLIMATE CHANGE WILL AFFECT **HUMANS**...

HALL OF MAN
AND WOMAN!

FINALLY, SOMETHING I CAN **CARE ABOUT**.
C'MON DAD, LET'S GO BACK TO THE **ANIMALS!**

...IT'S A BIT **HARD TO SAY** BECAUSE WE'RE SO GOOD AT **ADAPTING.**

WHAT IF **OCEAN ACIDIFICATION** DESTROYS MARINE ECOSYSTEMS?
WE'LL JUST EAT **FARMED FISH!**

BUT WHAT WOULD YOU **FEED** THE FARMED FISH?
CORN!

BUT HOW ARE YOU GOING TO **GROW ALL THAT CORN?**
GENETIC MODIFICATION!

BUT THEN THE FISH WILL TASTE LESS LIKE FISH AND MORE LIKE **CORN.**
DON'T WORRY, MY **TASTE BUDS** WILL **ADAPT.**

OF COURSE **ADAPTATION** WON'T NECESSARILY BE **EASY FOR ANYONE.**

THIS IS A THREAT TO MY **QUALITY OF LIFE!**

THIS IS A THREAT TO MY **ACTUAL LIFE.**

BUT IT WILL DEFINITELY BE **EASIER FOR THE WEALTHY**...

Homo sapiens opulentus

...THAN **FOR THE POOR.**

Homo sapiens penuriosus

YOU CAN SEE THE **GAP BETWEEN RICH AND POOR** WHEN YOU LOOK AT **SEA LEVEL RISE**...

OH, NO!
MY **SUMMER HOME!**

OH, NO!
MY **FARM!**

...**HEAT WAVES**...

AT LEAST WE'VE GOT **A/C** AND **ICED MARGARITAS!**

AT LEAST IT WON'T BE SO **COLD** IN THE **WINTER**.

...**EXTREME WEATHER**...

FLASH FLOODS?
DUST STORMS?
THIS 4X4 **HANDLES IT ALL**.

OUR TIN ROOF DOUBLES AS A **FLOTATION DEVICE**.

...HIGHER FOOD PRICES...
$10 FOR A MOLTO FRAPPUCCINO?
GUESS I WON'T BUY ANOTHER UNTIL TOMORROW.
$10 FOR A BAG OF RICE?
GUESS I WON'T BUY ANOTHER UNTIL NEXT MONTH.
...AND DISEASE OUTBREAKS.
AND THE IMPACT ON THE POOR IS ESPECIALLY UNFAIR SINCE THEY HAVEN'T CONTRIBUTED MUCH TO THE PROBLEM.

MY CARBON FOOTPRINT IS MORE LIKE A TOEPRINT.

FORTUNATELY, **ECONOMIC GROWTH** WILL LIKELY MEAN **MORE RICH PEOPLE** AND **FEWER POOR PEOPLE** IN THE DECADES AHEAD.

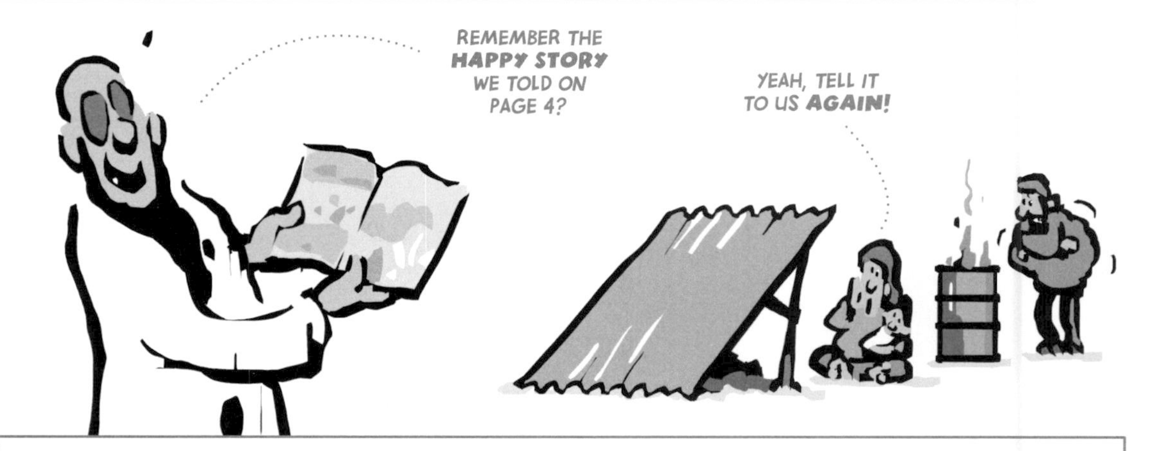
REMEMBER THE HAPPY STORY WE TOLD ON PAGE 4?
YEAH, TELL IT TO US AGAIN!

BUT EVEN THOSE WHO **BENEFIT** FROM ECONOMIC GROWTH WILL FACE **BIG CHALLENGES**...

I MISS ALL THAT BIODIVERSITY.
I MISS BEACHES.
WHADDYA MEAN I CAN'T EAT MONEY?

...AND THERE'S A CHANCE THAT EVEN THE RICH WILL BE **OVERWHELMED**.

ESPECIALLY TOWARDS THE **END OF THE CENTURY** AND BEYOND.

CHAPTER 10
BEYOND 2100

THE WORLD WILL CONTINUE BEYOND 2100, YOU KNOW.
AT LEAST WE HOPE SO!

SO FAR WE'VE FOCUSED ON THE IMPACTS OF CLIMATE CHANGE **IN THIS CENTURY.**

4°C
I DON'T **LIKE** THIS BOOK!

IN OTHER WORDS, WE'VE FOCUSED ON HOW CLIMATE CHANGE WILL AFFECT **US**...

...AND OUR **CHILDREN** AND **GRANDCHILDREN.**

BUT NOW WE'RE GOING TO LOOK AT WHAT MIGHT HAPPEN IN THE **EXTRA LONG TERM.**

LIKE TO **OUR** GREAT, GREAT, GREAT, GREAT GRANDCHILDREN.

UNFORTUNATELY, EVEN IF WE WENT **COLD TURKEY** AND STOPPED EMITTING GREENHOUSE GASES **TODAY**...

JUST MARRIED
STOP!
PARTY'S OVER.
STOP

...HUMAN-INDUCED CLIMATE CHANGE WOULD **CONTINUE BEYOND 2100.**

I SAID "**STOP**," BY ZORG!
JUST MARRIED

IN FACT, SEA LEVELS (AND MAYBE ALSO TEMPERATURES) WOULD **CONTINUE TO RISE FOR HUNDREDS OF YEARS.**

THE RISE WOULD JUST BE **SLOWER** THAN IF WE KEEP POLLUTING.
BUT **WHY?**

ONE REASON CLIMATE IMPACTS **GO ON AND ON** IS THAT CO_2 IS A **LONG–LIVED GAS.**

LONG LIVE CO_2!
CO_2
CO_2
CO_2
CO_2

WELL, THAT'S NOT REALLY **WHAT WE MEANT.**

ABOUT **HALF** OF THE CO_2 WE'RE EMITTING NOW WILL **STILL BE WARMING THE PLANET IN 100 YEARS...**

CO_2
CO_2
NYAA NYAA!

MOST OF THE REST WILL HAVE **DISSOLVED IN THE OCEANS...**
...MAKING THEM **MORE ACIDIC.**

...AND ABOUT **HALF OF THAT AMOUNT** WILL STILL BE AT IT IN **A THOUSAND YEARS.**

CO_2

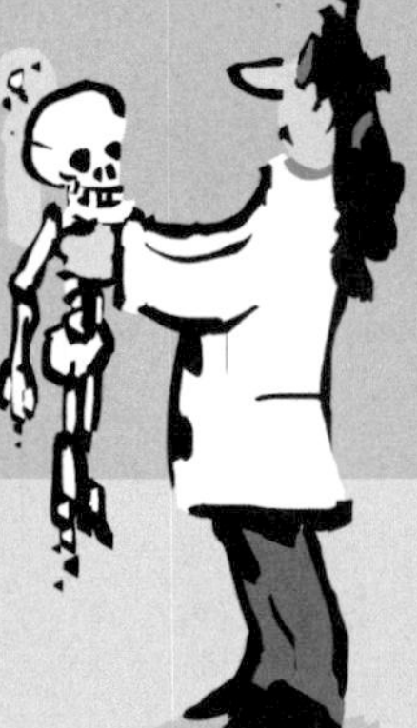
AS FAR AS HUMANS ARE CONCERNED, THAT'S PRETTY MUCH **FOREVER.**

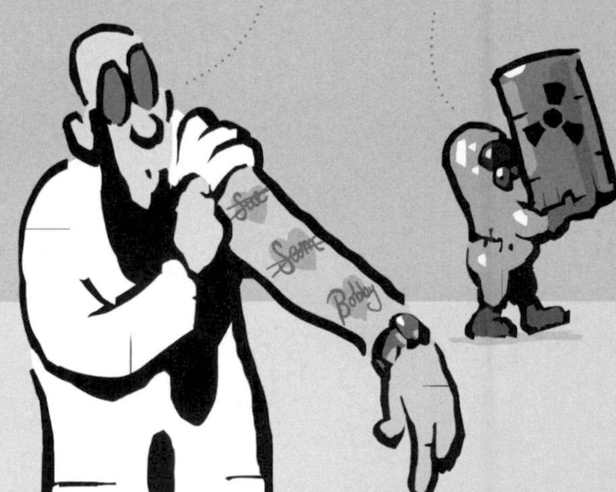
LIKE **TATTOOS...**
...OR **NUCLEAR WASTE.**
Bobby

ANOTHER REASON IS THAT **THE EARTH'S CLIMATE SYSTEM IS SLUGGISH.**

THIS MEANS THAT THE PLANET DOESN'T RESPOND **INSTANTANEOUSLY** TO CHANGES IN ENERGY BALANCE.

OUCH!

INSTEAD IT RESPONDS **GRADUALLY...**

DUDE!
OUCH!

... WHICH IN THIS CASE CAN MEAN **HUNDREDS** IF NOT **THOUSANDS OF YEARS.**

OUCH!

ONE ESPECIALLY **SLUGGISH** PART OF THE EARTH'S CLIMATE SYSTEM IS
SEA LEVEL RISE.

OK CLASS, TODAY WE'RE GOING TO WATCH **ICE MELT.**
ZZZZZZZ

WE SAW IN CHAPTER 8 THAT SEAS ARE LIKELY TO RISE ABOUT **2 FEET BY 2100.**

BUT SEAS WILL CONTINUE TO RISE FOR **SEVERAL CENTURIES BEYOND THAT...**

LOW TIDE IN 2200...

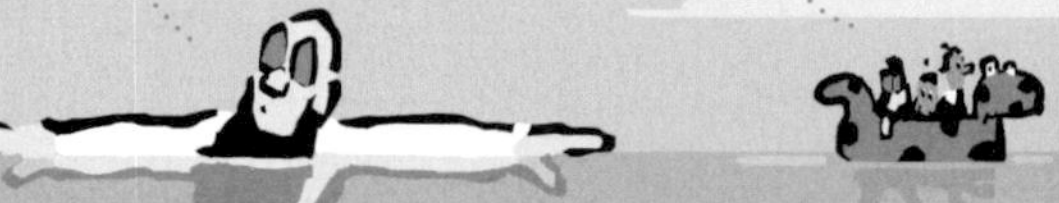
...MIGHT EQUAL **HIGH TIDE** IN 2100.

...PARTLY BECAUSE OF CONTINUING **THERMAL EXPANSION...**

WE LEARNED ABOUT THAT ON PAGE 92.

...AND PARTLY BECAUSE ONCE GLACIERS **START** TO MELT, IT'S HARD TO TELL WHEN THEY'LL **STOP.**

SEA LEVEL RISE IS ESPECIALLY PROBLEMATIC BECAUSE SO MANY PEOPLE LIVE **NEAR THE COAST.**

REAL ESTATE IS ALL ABOUT LOCATION, LOCATION, LOCATION.
Ocean View
Sea Side
Water Logged

IF THE ENTIRE **GREENLAND ICE SHEET** MELTS, THE SEAS WILL RISE BY ABOUT **20 FEET.**

GOODBYE, NEW ORLEANS.

GOODBYE, SHANGHAI.

IF THE **WEST ANTARCTIC ICE SHEET** MELTS, THEY'D RISE **ANOTHER 20 FEET.**

GOODBYE, FLORIDA.

GOODBYE, KOLKATA.

BUT REMEMBER, MELTING MILE-THICK ICE SHEETS WOULD TAKE **MANY HUNDREDS OF YEARS...**

...AND **WHO KNOWS** WHAT THE WORLD WILL LOOK LIKE IN **HUNDREDS OF YEARS?**

THE WORLD POPULATION COULD BE 2 BILLION...
...OR 36 BILLION.

WE COULD DISCOVER TECHNOGICAL MIRACLES...
...OR RETURN TO THE STONE AGE.

WE COULD HAVE UNDERWATER CITIES...
...OR WIND UP SWIMMING WITH THE FISHES.

TO SEE HOW HARD IT IS TO LOOK **BEYOND 2100**, IMAGINE SOMEONE IN **1900**...

HOWDY, STRANGER!

...TRYING TO ANTICIPATE **TODAY'S WORLD.**

ANTIBIOTICS?
COMPUTERS?
SATELLITE–GUIDED TRACTORS?
WHAT KIND OF NONSENSE ARE YOU TALKING?

INDEED, WHEN THE SWEDISH CHEMIST **ARRHENIUS** FIRST STUDIED CLIMATE CHANGE **100 YEARS AGO**...

...HE THOUGHT THE RESULTS WOULD BE **GOOD.**

"THE INCREASING PERCENTAGE OF [CO_2] IN THE ATMOSPHERE...

...WILL BRING FORTH MUCH MORE ABUNDANT CROPS THAN AT PRESENT."

OF COURSE, HE LIVED IN **SWEDEN.**

BRRRR.

ALL OF THIS JUST SHOWS HOW **TOUGH** IT IS TO **MAKE PREDICTIONS**...

...ESPECIALLY ABOUT THE **FUTURE!**

IN 100 YEARS WE'LL ALL BE BETTER OFF BECAUSE OF ECONOMIC GROWTH...

...UNLESS WE'RE WORSE OFF BECAUSE OF ENVIRONMENTAL DEGRADATION.

IN SUM, ALTHOUGH THE **GENERAL TRAJECTORY** FOR THE EARTH'S CLIMATE IS PRETTY CLEAR...

I'M JUST GETTING **WARMED UP.**

...THE **LONG TIME SPAN** ASSOCIATED WITH GLOBAL WARMING...

THE CHOICES **YOU MAKE** WILL HAVE CONSEQUENCES FOR YOUR **GREAT, GREAT, GREAT GRANDCHILDREN.**

...IS ONE OF THE **TWO MAJOR ISSUES** THAT ADD **UNCERTAINTY** TO THE PROBLEM OF CLIMATE CHANGE.

WHAT'S THE **OTHER** MAJOR ISSUE?

UNCERTAINTY.

CHAPTER 11
UNCERTAINTY

WELL, IF YOU WANT TO BE **CERTAIN**...
...WE'LL JUST HAVE TO **WAIT AND SEE.**

IN THIS CHAPTER WE'RE GOING TO TAKE A BREAK FROM LIKELY OUTCOMES...
TODAY WAS THE USUAL BLAH, BLAH, BLAH.

...AND THINK FOR A BIT ABOUT UNLIKELY OUTCOMES.

THE CEO GOT FIRED AND BURNED DOWN THE BUILDING!

CONGRATULATIONS!

CONGRATULATIONS!

I JUST GOT IN A FENDER BENDER...

...WITH AN ALIEN SPACECRAFT.

WHEN IT COMES TO CLIMATE, **UNLIKELY OUTCOMES** ARE, WELL, **UNLIKELY.**

COULD ALL OF GREENLAND MELT IN MY LIFETIME?
THAT WOULD BE EXCEPTIONALLY UNLIKELY.

BUT NO MATTER HOW YOU **MEASURE THE ODDS...**

...UNLIKELY DOESN'T MEAN **IMPOSSIBLE.**

THESE ODDS ARE LESS THAN 1 IN 3.
UNLIKELY

THESE ODDS ARE LESS THAN 1 IN 10.
VERY UNLIKELY

THESE ODDS ARE LESS THAN 1 IN 100.
EXCEPTIONALLY UNLIKELY

THESE ODDS ARE LESS THAN 1 IN 100,000,000.
EXTRA SUPER DUPER UNLIKELY
DUDE, I WON THE LOTTERY!

WE CAN SEE THIS BY LOOKING AT **PAST PREDICTIONS** MADE BY CLIMATE SCIENTISTS.

HEY, NOBODY'S PERFECT.

WHILE MANY SCIENTIFIC PREDICTIONS ABOUT CLIMATE CHANGE HAVE **HIT THE TARGET...**

WE THOUGHT **TEMPERATURES** WOULD RISE THIS MUCH...
...**AND THEY DID.**

WE THOUGHT **SEAS** WOULD RISE THIS MUCH...
...**AND THEY DID.**

...OTHERS HAVE **MISSED...**

...BOTH FOR **BETTER...**

WE THOUGHT **METHANE** WOULD RISE THIS MUCH...
...BUT IT ROSE **LESS.**

...AND FOR **WORSE.**

WE THOUGHT **ARCTIC SEA ICE** WOULD MELT...
...BUT NOT **THIS FAST.**

IN OTHER WORDS, SOME SCENARIOS SCIENTISTS THOUGHT WERE **UNLIKELY...**

...ACTUALLY CAME TO PASS.

XYZ IS VERY UNLIKELY.
NEWS FLASH:
XYZ JUST HAPPENED!
XYZ

MADDENING THOUGH IT IS, THIS SORT OF UNCERTAINTY IS **UNAVOIDABLE.**

TOUGH LUCK.

SOME OF THE **UNCERTAINTY** STEMS FROM ALL THE **POSITIVE FEEDBACK LOOPS** IN THE CLIMATE SYSTEM.

SCREEEEEEEEEEEEEEEEEE...
LIKE THE WATER VAPOR EFFECT...
...AND THE ICE-ALBEDO EFFECT.
HEY DUMMY, DON'T PUT THE MICROPHONE IN FRONT OF THE SPEAKER!

AND TO MAKE THINGS EVEN MORE DIFFICULT, THERE ARE ALSO **NEGATIVE FEEDBACK LOOPS.**

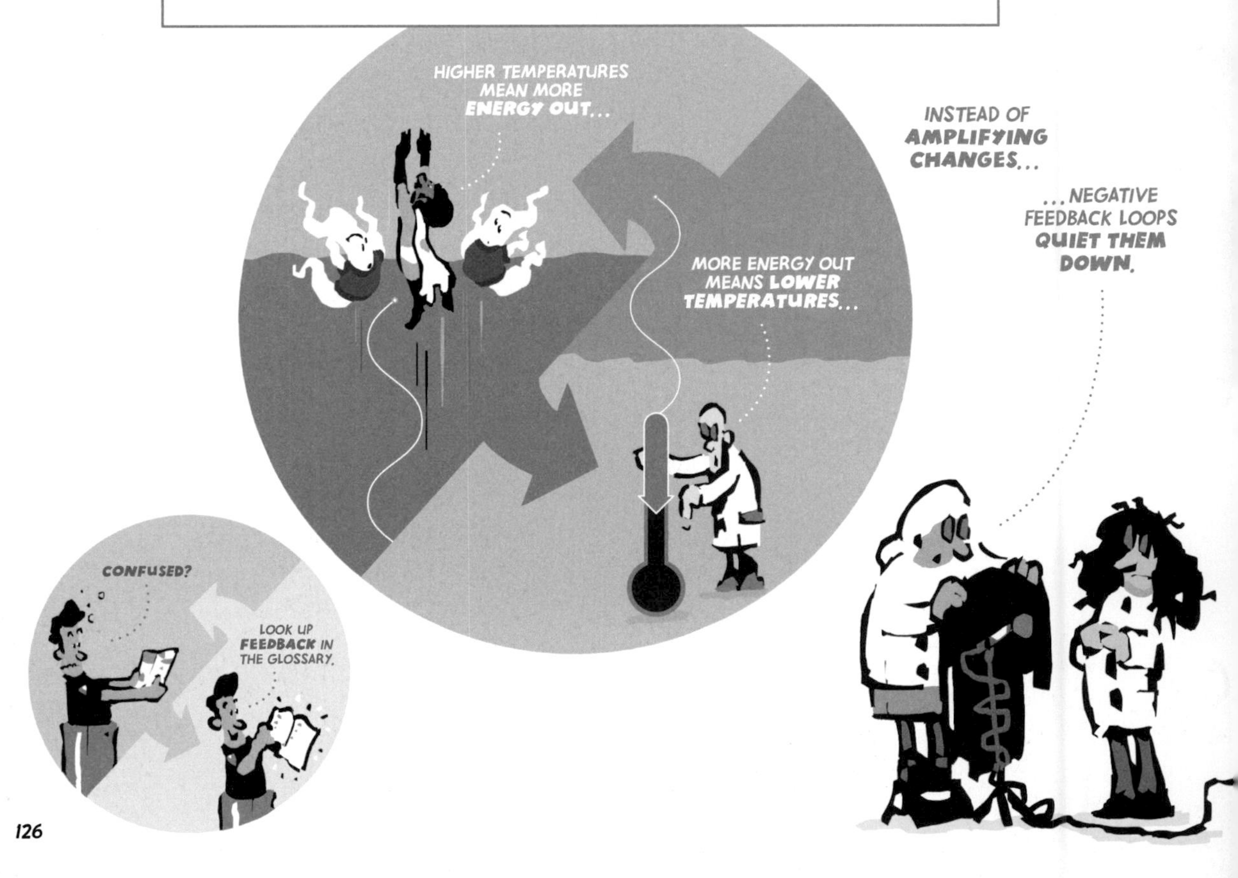

HIGHER TEMPERATURES MEAN MORE ENERGY OUT...
MORE ENERGY OUT MEANS LOWER TEMPERATURES...
INSTEAD OF AMPLIFYING CHANGES...
...NEGATIVE FEEDBACK LOOPS QUIET THEM DOWN.
CONFUSED?
LOOK UP FEEDBACK IN THE GLOSSARY.

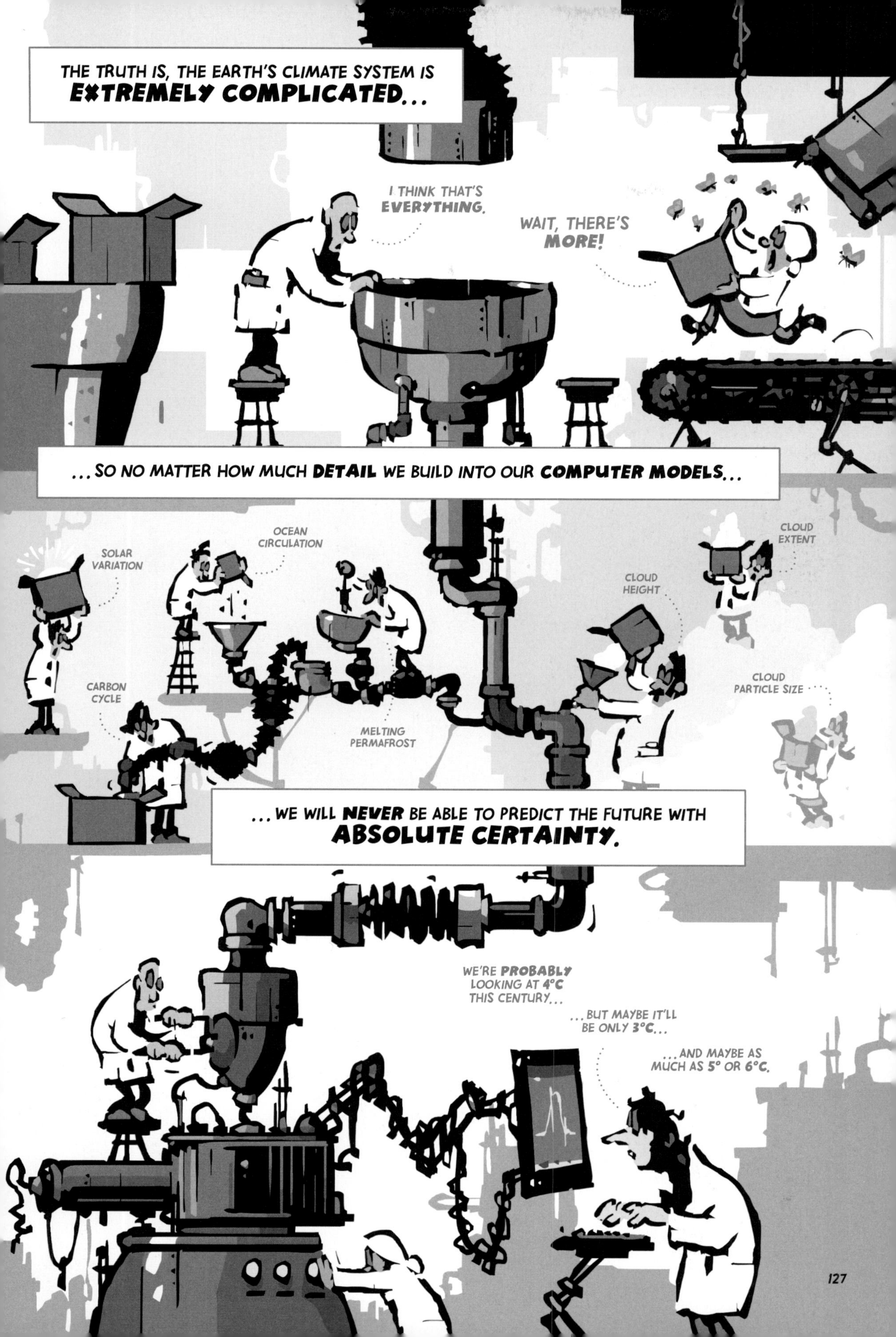
THE TRUTH IS, THE EARTH'S CLIMATE SYSTEM IS EXTREMELY COMPLICATED...
I THINK THAT'S EVERYTHING.
WAIT, THERE'S MORE!
...SO NO MATTER HOW MUCH DETAIL WE BUILD INTO OUR COMPUTER MODELS...
SOLAR VARIATION
OCEAN CIRCULATION
CLOUD EXTENT
CLOUD HEIGHT
CARBON CYCLE
MELTING PERMAFROST
CLOUD PARTICLE SIZE
...WE WILL NEVER BE ABLE TO PREDICT THE FUTURE WITH ABSOLUTE CERTAINTY.
WE'RE PROBABLY LOOKING AT 4°C THIS CENTURY...
...BUT MAYBE IT'LL BE ONLY 3°C...
...AND MAYBE AS MUCH AS 5° OR 6°C.

AND THAT LEAVES US WITH A **CONUNDRUM.**

IF REALITY TURNS OUT TO BE **BETTER** THAN WE'D THOUGHT...

CLIMATE SENSITIVITY IS **LOW**...
...ECOSYSTEMS ARE **RESILIENT**...
...PEOPLE FIND WAYS TO **ADAPT**.
WHEW, WE **DODGED A BULLET** BACK THERE.
3°C

...THEN BUSINESS AS USUAL MIGHT **NOT BE SO BAD.**

THANKS TO RAPID ECONOMIC GROWTH, WE ALL GET **MORE CAKE**...
...**AND** THERE'S NOT TOO MUCH EXTRA SUFFERING FROM CLIMATE CHANGE.

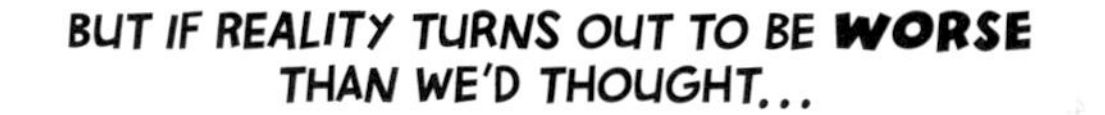
BUT IF REALITY TURNS OUT TO BE **WORSE** THAN WE'D THOUGHT...
CLIMATE SENSITIVITY IS **HIGH**...
...ICE SHEETS **DISINTEGRATE**...
...THE AMAZON **BURNS UP**...
...BREAD BASKETS BECOME **DUST BOWLS**.
6°C
...THEN BUSINESS AS USUAL COULD BE **CATASTROPHIC**.
INSTEAD OF **MORE CAKE**...
...ALL WE GOT WAS **TRAGEDY**.

ALL THIS TALK ABOUT **POTENTIAL CATASTROPHES...**

...SHOULD MAKE YOU THINK ABOUT **INSURANCE.**

I'M **SCARED.**
WHAT CAN I **DO?**
TALK TO MY SISTER...

...SHE'S A BROKER AT **AETNA.**

FOR EXAMPLE, WHILE IT'S **UNLIKELY** THAT YOU'LL GET INTO A **HORRIFIC ACCIDENT...**

ESPECIALLY WITH AN **ALIEN.**

...OR BECOME **DEATHLY ILL...**

MEDIC!
IT'S ANOTHER CASE OF **STOCHASTIC ANXIETY DISORDER.**

...IT'S A GOOD IDEA TO **BUY INSURANCE,** JUST IN CASE.

INSURANCE GIVES YOU A WAY TO **PAY A BIT MORE NOW...**
...TO AVOID **TOTAL DISASTER IN THE FUTURE.**
BAD TURNS INTO **A BIT LESS BAD.**

SIMILARLY, ALTHOUGH IT'S UNLIKELY THAT CLIMATE IMPACTS WILL BE CATASTROPHIC...
THIS REMINDS ME OF VENUS.
I THOUGHT MEN WERE FROM MARS.
...IT'S A GOOD IDEA TO BUY INSURANCE, JUST IN CASE.

I LIKE **SCIENCE FICTION**.

BUT WE DON'T WANT OUR **KIDS** TO **LIVE THROUGH IT**.

NOW, YOU CAN BUY INSURANCE FOR ALL KINDS OF THINGS...
I'VE GOT LIFE INSURANCE...
...PET INSURANCE...
...HOME INSURANCE...
...FLOOD INSURANCE...
...EVEN ALIEN ABDUCTION INSURANCE.
...BUT HOW DO YOU BUY INSURANCE FOR THE WHOLE PLANET?

IT TURNS OUT THAT OUR
BEST INSURANCE POLICY...

...IS TO **REDUCE EMISSIONS** OF CO_2
AND OTHER GREENHOUSE GASES.

THE MORE WE REDUCE **NOW**...
...THE LESS LIKELY WE'LL **MEET CATASTROPHE**.

AND WE'LL LEARN MORE ABOUT THAT IN THE **NEXT PART OF THIS BOOK**.

IF WE GIVE UP A SMALL **PIECE OF CAKE**...
...WE CAN GET **PEACE OF MIND**.

PART THREE
ACTIONS

CHAPTER 12
THE TRAGEDY OF THE COMMONS

SOMEBODY'S GOT TO DO SOMETHING...
...RIGHT?

FOSSIL FUELS ARE A **KEY INGREDIENT** IN MODERN ECONOMIES...

EVERYONE USES THEM.
YOU.
ME.
HER.
HIM.
ALL OF US.
China
India
Rest of Asia
Rich World
Everyone Else

...SO IT'S **NOT GOING TO BE EASY** TO **SLOW DOWN** THE GROWTH IN FOSSIL FUEL CONSUMPTION.

WE WANT TO **USE THEM TOO!**
DUH!

NEVERTHELESS, WHEN YOU COMPARE THE **BENEFITS** OF **REDUCING CO_2 EMISSIONS**...

...WITH THE **COSTS**...

A MORE **STABLE CLIMATE**.

LOWER RISK OF A **CLIMATE CATASTROPHE**.

LESS **SEA LEVEL RISE**.

ECONOMIC GROWTH IN **RENEWABLE ENERGY**.

FEWER **EXTINCTIONS**.

OPPORTUNITIES TO **REDUCE EXISTING TAXES**.

HIGHER PRICES FOR **GASOLINE**...

...AND **COAL** AND **NATURAL GAS**...

...AND, THEREFORE, **ELECTRICITY**.

ECONOMIC LOSSES IN **COAL MINES**.

...MOST PEOPLE CONCLUDE THAT IT **MAKES GOOD SENSE** TO TRY.

WE NEED SOME **INSURANCE!**
DUH!

BUT JUST BECAUSE TAKING ACTION **MAKES GOOD SENSE**...

I'M TRYING TO STOP SMOKING...

...DOESN'T MEAN IT WILL **ACTUALLY HAPPEN.**

...BUT I JUST CAN'T!

THAT'S A PROBLEM FOR **INDIVIDUALS**, BUT IT'S AN EVEN BIGGER PROBLEM FOR **GROUPS**...

WE'RE TRYING TO REDUCE CO2...

...BUT WE JUST CAN'T!

...BECAUSE OF THE **TRAGEDY OF THE COMMONS.**

INDIVIDUAL SELF-INTEREST...
...CAN LEAD TO BAD OUTCOMES FOR THE GROUP AS A WHOLE.

THE IDEA BEHIND THE TRAGEDY OF THE COMMONS GOES BACK AT LEAST AS FAR AS **ARISTOTLE...**

"THAT WHICH IS **COMMON TO THE GREATEST NUMBER...**
...HAS THE **LEAST CARE** BESTOWED UPON IT."

...AND, FOR ALL WE KNOW, HE WAS INSPIRED BY **DOG POOP.**

#%*@!

SELF-INTEREST LEADS PEOPLE TO PICK UP AFTER THEIR **OWN DOG** IN THEIR **OWN BACKYARD...**

...BUT LOTS OF PEOPLE DON'T PICK UP AFTER THEIR DOG IN **PUBLIC PLACES.**

"EVERYONE THINKS CHIEFLY OF **THEIR OWN...**
...HARDLY AT ALL OF THE **COMMON INTEREST.**"

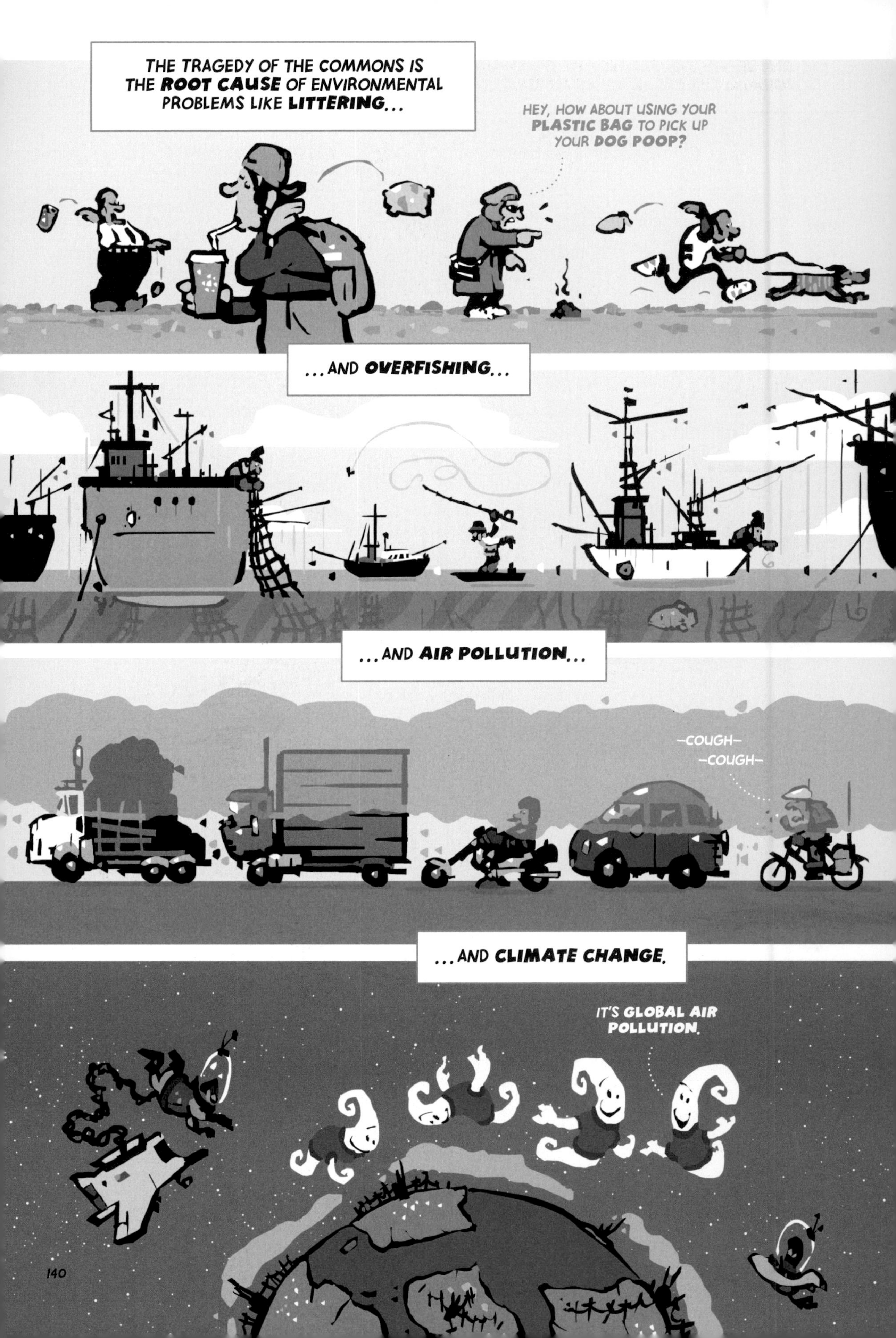
THE TRAGEDY OF THE COMMONS IS THE ROOT CAUSE OF ENVIRONMENTAL PROBLEMS LIKE LITTERING...
HEY, HOW ABOUT USING YOUR PLASTIC BAG TO PICK UP YOUR DOG POOP?
...AND OVERFISHING...
...AND AIR POLLUTION...
-COUGH-
-COUGH-
...AND CLIMATE CHANGE.
IT'S GLOBAL AIR POLLUTION.

YOU CAN SEE THE TRAGEDY OF THE COMMONS IN YOUR OWN **PERSONAL DECISIONS**...

SURE I'LL GENERATE CARBON EMISSIONS IF I FLY TO HAWAII...
...BUT I REALLY WANT TO GO TO HAWAII!

...AND IN DECISIONS MADE BY ENTIRE COUNTRIES WHEN THEY FAIL TO SIGN CLIMATE AGREEMENTS...
AMERICAN SELFISH JERK!
...OR FAIL TO LIVE UP TO AGREEMENTS THEY ALREADY SIGNED...
CANADIAN HYPOCRITE!

...OR FAIL TO DO MUCH BEYOND **BLAMING EVERYONE ELSE.**

FORTUNATELY, THERE ARE **TWO PIECES** OF **GOOD NEWS.**

ONE IS THAT **SELF-INTEREST IS NOT ALL BAD.**

"IT IS NOT FROM THE BENEVOLENCE OF THE BUTCHER, THE BREWER, OR THE BAKER THAT WE EXPECT OUR DINNER...
...BUT FROM THEIR REGARD TO THEIR OWN INTEREST."

FOR EXAMPLE, SOME COUNTRIES HAVE **STRONG INCENTIVES** FOR **ADAPTATION...**

IF WE DON'T ADAPT, WE'LL BE FLOODED...
...SO LET'S BUILD A SEAWALL!

...AND SOMETIMES **SELF-INTEREST** HAPPENS TO COINCIDE WITH **EMISSIONS REDUCTIONS.**

WE WANT TO CLEAN UP LOCAL POLLUTION...
...AND IMPROVE ENERGY SECURITY...
...AND BURNING LESS FOSSIL FUELS WILL HELP US.

THE OTHER PIECE OF GOOD NEWS IS THAT THE TRAGEDY OF THE COMMONS IS **NOT INEVITABLE.**

WE DON'T **HAVE** TO FIGHT.

HISTORY SHOWS THAT **INTERNATIONAL AGREEMENTS** ARE IN FACT **POSSIBLE**...

THE **MONTREAL PROTOCOL** HELPED PROTECT THE OZONE LAYER.
AND COUNTRIES HAVE ALSO COOPERATED ON **FREE TRADE** AGREEMENTS...
...AND **FISHING TREATIES**...
...AND **ARMS CONTROL** AGREEMENTS.

...AND IN CHAPTER 14 WE'LL SEE WHAT A **CLIMATE AGREEMENT** MIGHT LOOK LIKE.

HOW ABOUT A GLOBAL AGREEMENT ON **CARBON TAXES?**
OR A BINDING **CAP AND TRADE** TREATY?

BUT TO OVERCOME THE TRAGEDY OF THE COMMONS, AN AGREEMENT NEEDS TO HAVE **TEETH**...

$500 Fine

CARBON SCOFFLAWS WILL FACE **INTERNATIONAL SANCTIONS!**

...AND PERHAPS EVEN AN **ELEMENT OF FORCE.**

FREEZE!

SIGN THIS CLIMATE AGREEMENT OR WE'LL IMPOSE **BORDER TAX ADJUSTMENTS** ON YOUR COUNTRY.
AND IF YOU DON'T UNDERSTAND THAT, **READ THE GLOSSARY!**

AND THE DIFFICULTY OF **GETTING EVERYONE TO AGREE**...

NO FAIR!
HER FINE SHOULD BE **BIGGER THAN MINE!**

NO FAIR!
THIS BURDENS MY COUNTRY **MORE THAN YOURS!**

...LEADS SOME FOLKS TO HOPE FOR A **MIRACLE CURE.**

TEFLON SHOES!
ROBOT POOPER–SCOOPERS!
GENETICALLY ENGINEERED **POOPLESS DOGS!**

CHAPTER 13
TECHNO–FIX

NO PROBLEM, I CAN FIX IT RIGHT UP!

SOME IDEAS FOR TACKLING CLIMATE CHANGE SOUND LIKE SCIENCE FICTION...
...LIKE FERTILIZING THE OCEANS WITH IRON...

THAT STIMULATES THE GROWTH OF PLANKTON, WHICH ABSORBS CO_2.
WE JUST NEED THE PLANKTON TO DIE AND SINK TO THE BOTTOM.
...SPRAYING SALTWATER INTO THE AIR...

THAT INCREASES THE EARTH'S ALBEDO BY CREATING CLOUDS.

WE JUST NEED AN ARMADA OF 30,000 SELF-GUIDED SHIPS.
...PUMPING SULFUR DIOXIDE INTO THE ATMOSPHERE...

SULFUR PARTICLES REFLECT INCOMING SUNLIGHT.
SO_2

WE JUST NEED SOME 18-MILE-LONG GARDEN HOSES.
...OR ENGINEERING GENETICALLY MODIFIED CARBON-EATING TREES.

COULD YOU ALSO DESIGN THEM TO GIVE ME DIRECTIONS?

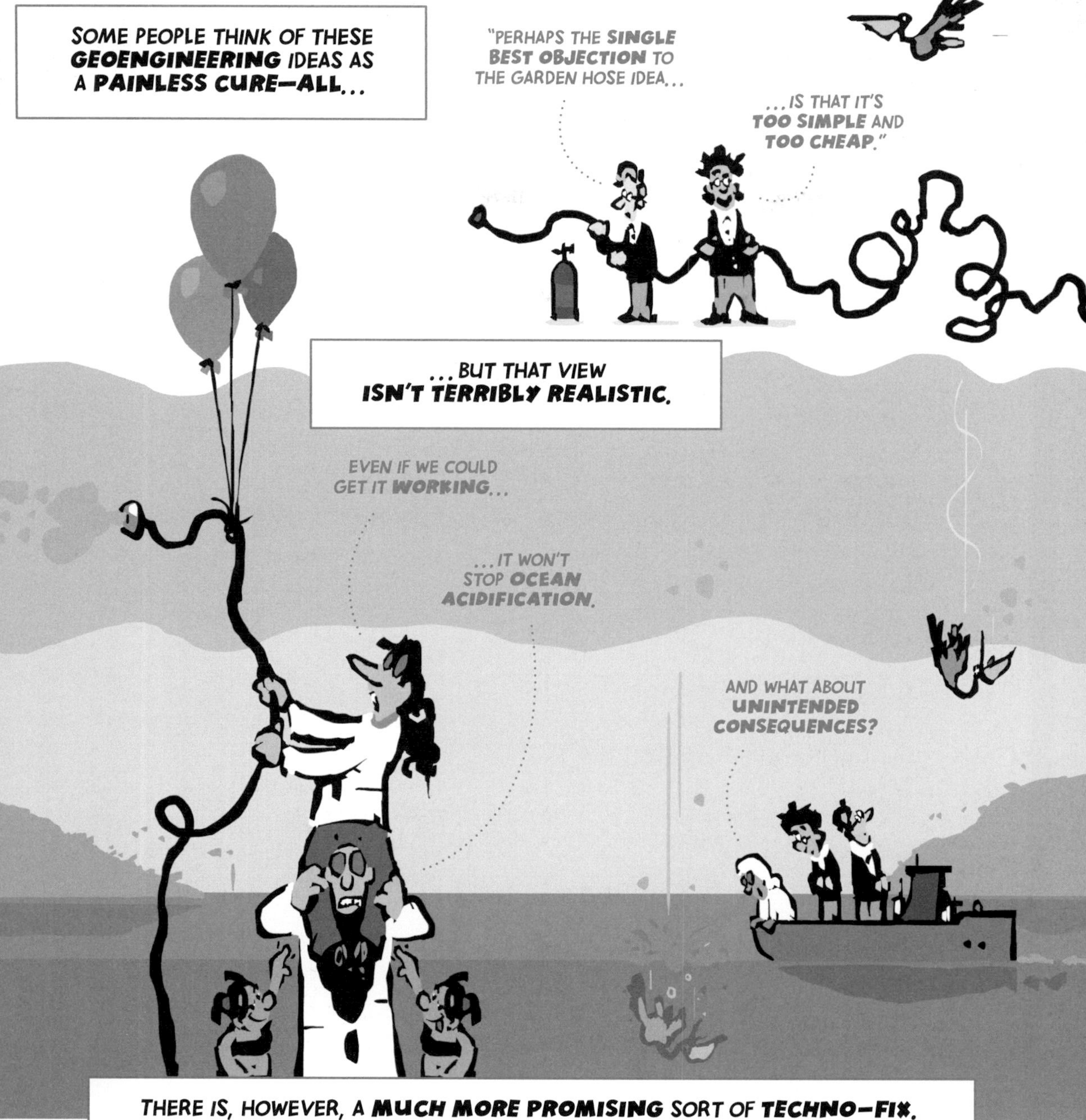
SOME PEOPLE THINK OF THESE GEOENGINEERING IDEAS AS A PAINLESS CURE-ALL...
"PERHAPS THE SINGLE BEST OBJECTION TO THE GARDEN HOSE IDEA...
...IS THAT IT'S TOO SIMPLE AND TOO CHEAP."
...BUT THAT VIEW ISN'T TERRIBLY REALISTIC.
EVEN IF WE COULD GET IT WORKING...
...IT WON'T STOP OCEAN ACIDIFICATION.
AND WHAT ABOUT UNINTENDED CONSEQUENCES?

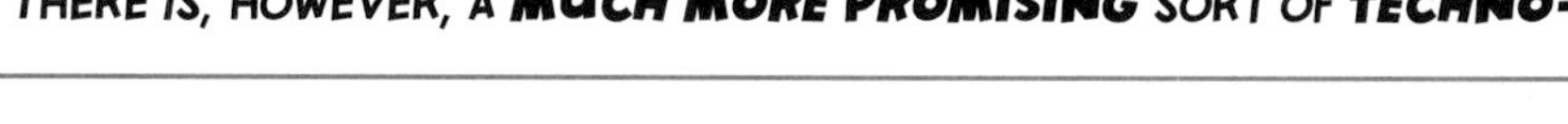
THERE IS, HOWEVER, A MUCH MORE PROMISING SORT OF TECHNO-FIX.

WHAT IS IT, SPACE MIRRORS?

NO, SILLY.
IT'S CHEAP, CLEAN ENERGY.

IF WE CAN DEVELOP **CLEAN ENERGY** THAT'S **CHEAPER THAN FOSSIL FUELS**...

ENERGY FROM FOSSIL FUELS IS CHEAP...
...BUT MY CLEAN ENERGY IS CHEAPER.

...THEN IT WILL BE IN EVERYONE'S **SELF-INTEREST** TO ABANDON COAL, OIL, AND NATURAL GAS.

NO TRAGEDY OF THE COMMONS.
NO PROBLEM!

AND CHEAP, CLEAN ENERGY WOULD NOT ONLY HELP **STABILIZE THE CLIMATE.**

TAKE THAT, BUSINESS AS USUAL!

IT WOULD ALSO **IMPROVE LIVING STANDARDS** FOR BILLIONS OF PEOPLE...

...AND CLEAN UP **LOCAL AIR POLLUTION.**

CHEAP ENERGY COULD BRING US REFRIGERATION...
...TRANSPORTATION...
...AND ACCESS TO THE INTERNET.

CLEAN ENERGY COULD BRING US BLUE SKIES.

SOUNDS **GREAT**, RIGHT?

SO...
...WHAT IS THIS CHEAP, CLEAN ENERGY?
GOOD QUESTION!

FROM A CLIMATE PERSPECTIVE, **ENERGY IS CLEAN** IF IT **DOESN'T INCREASE ATMOSPHERIC CO_2 CONCENTRATIONS.**

FOSSIL FUELS PUSH UP THE KEELING CURVE.

CLEAN ENERGY SOURCES DON'T.

THIS INCLUDES **RENEWABLE** ENERGY SOURCES LIKE **SOLAR POWER**...

...**HYDROPOWER**...

...**WIND POWER**...

...**GEOTHERMAL POWER**...

...AND **TIDAL POWER.**

IT **ALSO** INCLUDES ALTERNATIVES LIKE **NUCLEAR POWER**...
FROM A **CLIMATE** PERSPECTIVE, I'M CLEAN.

...AND SOME **BIOFUELS**...
GROWING BIOFUELS **CONSUMES CO_2**...
...SO BURNING THEM CAN BE **CARBON NEUTRAL.**

...AS WELL AS PROPOSALS SUCH AS **CARBON CAPTURE AND STORAGE.**
IF WE BURN FOSSIL FUELS AND PUMP THE CO_2 **UNDERGROUND**...

...IT **WON'T END UP IN THE ATMOSPHERE.**

YOU CAN THINK OF **ENERGY EFFICIENCY** AS A FORM OF CLEAN ENERGY, TOO.
IF WE FIND WAYS TO **REDUCE OUR ENERGY CONSUMPTION**...
...WE CAN GENERATE **"NEGAWATTS."**

THERE ARE ALL SORTS OF **CHALLENGES** FACING THESE CLEAN ENERGY OPTIONS...

THE SUN **DOESN'T ALWAYS SHINE.**
THE WIND **DOESN'T ALWAYS BLOW.**

...BUT THE **MAIN CHALLENGE** FACING ALL OF THEM CAN BE SUMMED UP IN **ONE WORD.**

MONEY

IT MIGHT **FEEL** LIKE **FOSSIL FUELS ARE EXPENSIVE.**

$4 A GALLON!
$180 A MONTH!

BUT FOSSIL FUELS ARE STILL USUALLY THE **CHEAPEST OPTION** FOR **BUYERS**...

...AND FOR **SELLERS.**

THAT'S WHY WE KEEP BUYING THEM.
THAT'S WHY WE KEEP SELLING THEM.

IN PART THAT'S BECAUSE OF ALL THE **INVESTMENTS** OVER THE PAST CENTURY...

IT COST ME A LOT TO BUILD THIS COAL PLANT...
...BUT NOW THAT IT'S BUILT MY OPERATING COSTS ARE LOW.

...AND BECAUSE OF ONGOING **FOSSIL FUEL SUBSIDIES** ESTIMATED AT **$400 BILLION** PER YEAR.

IN ORDER TO MAKE CLEAN ENERGY **MORE COMPETITIVE...**

...MANY PEOPLE WANT THE **GOVERNMENT** TO **SUBSIDIZE RENEWABLES.**

LET'S FUND RESEARCH AND DEVELOPMENT.

LET'S PAY HOMEOWNERS TO PUT UP SOLAR PANELS.

LET'S REQUIRE UTILITIES TO USE AT LEAST 25% RENEWABLES.

LET'S PROVIDE A TAX CREDIT FOR WIND FARMS.

MANY ECONOMISTS, HOWEVER, EMPHASIZE THAT FOSSIL FUELS ARE **NOT ACTUALLY AS CHEAP AS THEY APPEAR...**

THEY'RE NOT CHEAP IN MY BOOK.

BUT THEY SHOULD BE EVEN MORE NOT CHEAP!

...BECAUSE THEIR PRICE DOESN'T INCLUDE THE **COSTS OF POLLUTION.**

THE PRICE YOU PAY DOESN'T INCLUDE THESE THINGS.
Global Warming
Health Impacts
Ocean Acidification

AND THINGS CERTAINLY WOULD **LOOK DIFFERENT...**

...IF WE TOOK STEPS TO **INTERNALIZE** THOSE **EXTERNAL COSTS.**

Ocean Acidification
Global Warming
Health Impacts
HEY!
YOU'RE BUYING THE FOSSIL FUELS...
...YOU SHOULD PAY THE TRUE COST.

THAT'S WHY MOST ECONOMISTS **AGREE**...

...THAT THE BEST WAY TO MAKE CLEAN ENERGY **MORE COMPETITIVE WITH FOSSIL FUELS**...

...IS TO **MAKE FOSSIL FUELS MORE EXPENSIVE.**

WE NEED TO LEVEL THE PLAYING FIELD.

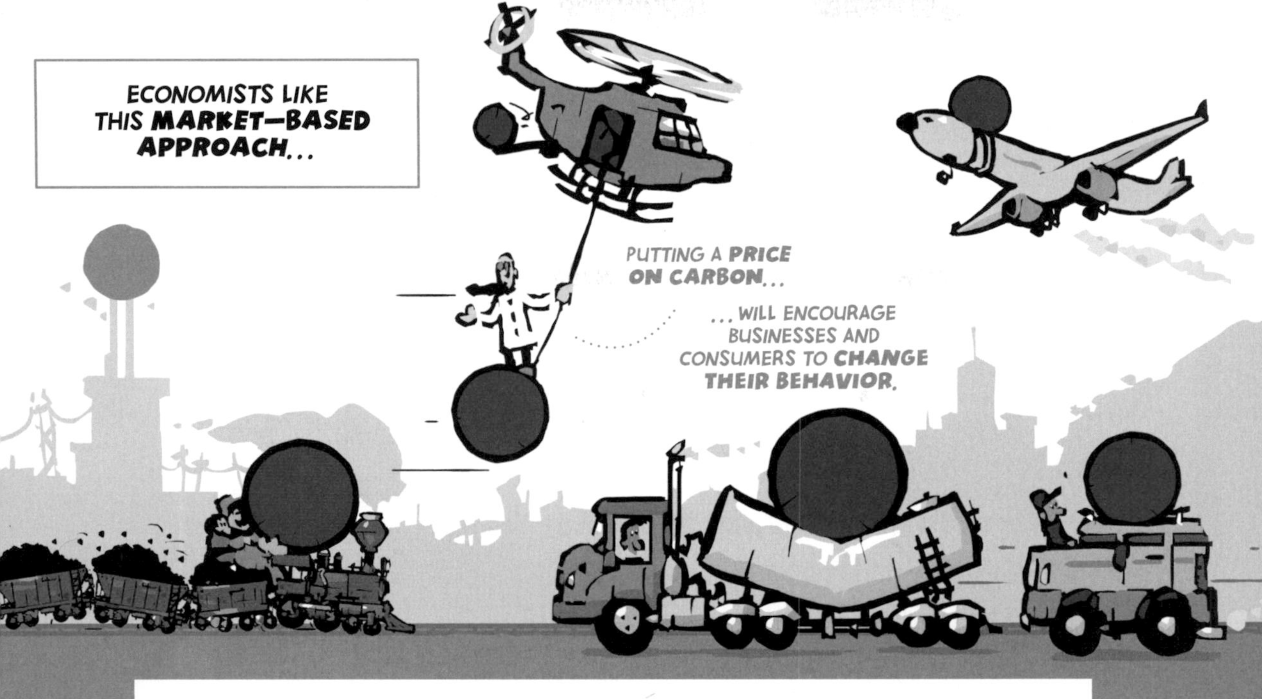
ECONOMISTS LIKE THIS **MARKET-BASED APPROACH...**
PUTTING A **PRICE ON CARBON...**
...WILL ENCOURAGE BUSINESSES AND CONSUMERS TO **CHANGE THEIR BEHAVIOR.**

...BECAUSE IT PROMISES TO BOTH **REDUCE THE CONSUMPTION OF DIRTY FUELS** IN THE **SHORT RUN...**

IT'S IN OUR **SELF-INTEREST** TO ABANDON YOU.

...AND **PROMOTE THE DEVELOPMENT OF NEW TECHNOLOGIES** IN THE **LONG RUN.**

IT'S IN YOUR **SELF-INTEREST** TO INVEST IN CLEAN ENERGY.

OF COURSE, THE IDEA OF MAKING FOSSIL FUELS MORE EXPENSIVE IS **CONTROVERSIAL.**

HOLD ON, YOU WANT **ME** TO PAY MORE FOR FOSSIL FUELS?

OF COURSE!

WHAT, DID YOU THINK POLLUTION JUST COMES FROM **BIG COMPANIES?**

IN PART THAT'S BECAUSE THE **LIGHT AT THE END OF THE TUNNEL...**

I CAN SEE **CHEAP, CLEAN ENERGY.**

JUST A **LITTLE BIT FURTHER.**

...IS AT THE **END OF A TUNNEL.**

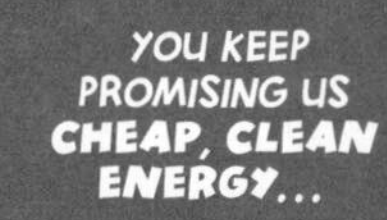
YOU KEEP PROMISING US **CHEAP, CLEAN ENERGY...**

...BUT SO FAR ALL YOU'VE DONE IS **MAKE DIRTY ENERGY MORE EXPENSIVE.**

SO WE'RE GOING TO SPEND THE NEXT CHAPTER PROVIDING SOME **ILLUMINATION.**

WE'VE GOT SOLUTIONS THAT MAKE SENSE IN BOTH THE **SHORT RUN...**

...AND THE **LONG RUN.**

CHAPTER 14
PUTTING A PRICE ON CARBON

WE CAN USE THE POWER OF CAPITALISM...
...AND THE TOOLS OF ECONOMICS...
...TO PROTECT THE ENVIRONMENT.

TACKLING CLIMATE CHANGE WILL PROBABLY REQUIRE ACTION BY **VOTERS** AND **GOVERNMENTS.**

WE'VE GOT TO **WORK TOGETHER** TO OVERCOME THE **TRAGEDY OF THE COMMONS.**

THE TWO MAIN POLICY OPTIONS ARE **DIRECT REGULATIONS...**

NEW CARS **MUST** AVERAGE 40 MILES A GALLON.

APPLIANCES **MUST** MEET NEW EFFICIENCY STANDARDS.

POWER COMPANIES **MUST** USE 25% RENEWABLES.

IF YOU **MUST** KNOW...
... THESE POLICIES ARE ALSO CALLED **COMMAND AND CONTROL.**

...AND **MARKET-BASED APPROACHES** LIKE **CARBON TAXES** AND **CAP AND TRADE SYSTEMS.**

THE WAY THESE POLICIES **REDUCE POLLUTION...**
...IS BY **MAKING POLLUTING EXPENSIVE.**

ONE DRAWBACK OF **DIRECT REGULATIONS** IS THAT THEY CAN BE PRETTY **HEAVY–HANDED.**

YOU **MUST** USE THIS KIND OF **LIGHT BULB**...
...AND THIS KIND OF **CAR**...
...AND THIS KIND OF **TOILET**...

ARRGH!

IN CONTRAST, MARKET–BASED APPROACHES OFFER MORE **FLEXIBILITY.**

WE'RE NOT GOING TO **FORCE** YOU TO DO **ANYTHING.**
WE'RE JUST GOING TO **MAKE POLLUTING EXPENSIVE**...
...AND LET YOU RESPOND **AS YOU SEE FIT.**
WELL, AT LEAST YOU'RE **NOT MICROMANAGING** ME.

THEY ALSO PROMOTE **CONSERVATION AND INNOVATION**...

I'M NOT GOING TO BUY THAT **HUMMER**...
...I'M GOING TO BUY THIS **SUPER–EFFICIENT HYBRID.**

...AND THEY DO ALL THIS BY **PUTTING A PRICE ON CARBON.**

CO_2
RIGHT NOW EMITTING CO_2 IS **FREE.**

THIS CHAPTER IS ABOUT HOW TO CHANGE THAT.

THE OBVIOUS WAY TO PUT A PRICE ON CARBON IS WITH A **CARBON TAX**...

A TAX?

NOW YOU'VE **GOT OUR ATTENTION.**

...LIKE THE ONE IN **BRITISH COLUMBIA.**

I DON'T LIKE TO **BRAG**, BUT WE HAVE THE **BEST CLIMATE POLICY IN THE WORLD.**

THE B.C. CARBON TAX APPLIES TO "UPSTREAM" BUSINESSES LIKE **OIL REFINERIES** AND **POWER PLANTS**...

THE TAX IS **$30 PER TON** OF CO2...
...EQUAL TO ABOUT **$0.30 PER GALLON** OF GASOLINE...
...OR **$0.03 PER KILOWATT-HOUR** OF COAL-FIRED POWER.

...BUT THOSE COMPANIES MOSTLY **PASS THE TAX ALONG TO CONSUMERS.**

THE TAX GIVES **US** AN INCENTIVE TO USE **LESS FOSSIL FUELS**...

...AND NOW **YOU** HAVE AN INCENTIVE, **TOO.**

SORRY, BUT THAT'S HOW MARKET FORCES WORK.

LUCKILY, THERE ARE **CLEVER WAYS** TO **REDUCE THESE BURDENS.**

THE CARBON TAX IN B.C. IS **ESPECIALLY CLEVER** BECAUSE THE GOVERNMENT USES THE **TAX REVENUE**...

LOOK AT THESE **BILLIONS** OF EXTRA DOLLARS, EH?

...TO **REDUCE EXISTING TAXES** ON JOBS, INCOME, AND INVESTMENT.

RAISING TAXES ON THINGS WE WANT **LESS OF**...

...ALLOWS US TO **LOWER TAXES** ON THINGS WE WANT **MORE OF**.

THERE ARE, OF COURSE, **OTHER WAYS** TO "RECYCLE" THE REVENUE FROM A CARBON TAX...

LET'S INVEST IN **R&D**...

...OR PROMOTE **ENERGY EFFICIENCY**...

...OR PAY DOWN THE **DEFICIT**...

...OR GIVE EVERYONE A **DIVIDEND CHECK**.

...BUT WHAT THEY ALL HAVE **IN COMMON** IS THAT THE REVENUE CAN BE PUT TO **GOOD USE**.

THOSE GOOD USES ARE ONE OF THE **SIDE BENEFITS** OF CARBON TAXES...
...IN ADDITION TO THE MAIN BENEFIT OF POTENTIALLY **SAVING THE PLANET**.

ANOTHER WAY TO PUT A PRICE ON CARBON IS WITH **CAP AND TRADE.**

LIKE THE EMISSIONS TRADING SYSTEM IN EUROPE...

...OR THE GLOBAL WARMING SOLUTIONS ACT IN CALIFORNIA.

THE WAY **POLITICIANS** TALK ABOUT CAP AND TRADE CAN MAKE IT SOUND LIKE **MAGIC...**

WE'RE GOING TO CAP EMISSIONS...
...AND LET PEOPLE MAKE MONEY FROM TRADING.
WHAT'S NOT TO LIKE?

...A SPECIAL KIND OF MAGIC THAT HAS **NOTHING IN COMMON WITH TAXES.**

TAXES ARE BAD...
...TRADING IS GOOD!

BUT **ECONOMISTS KNOW** THAT CAP AND TRADE AND CARBON TAXES ARE ACTUALLY **QUITE SIMILAR.**

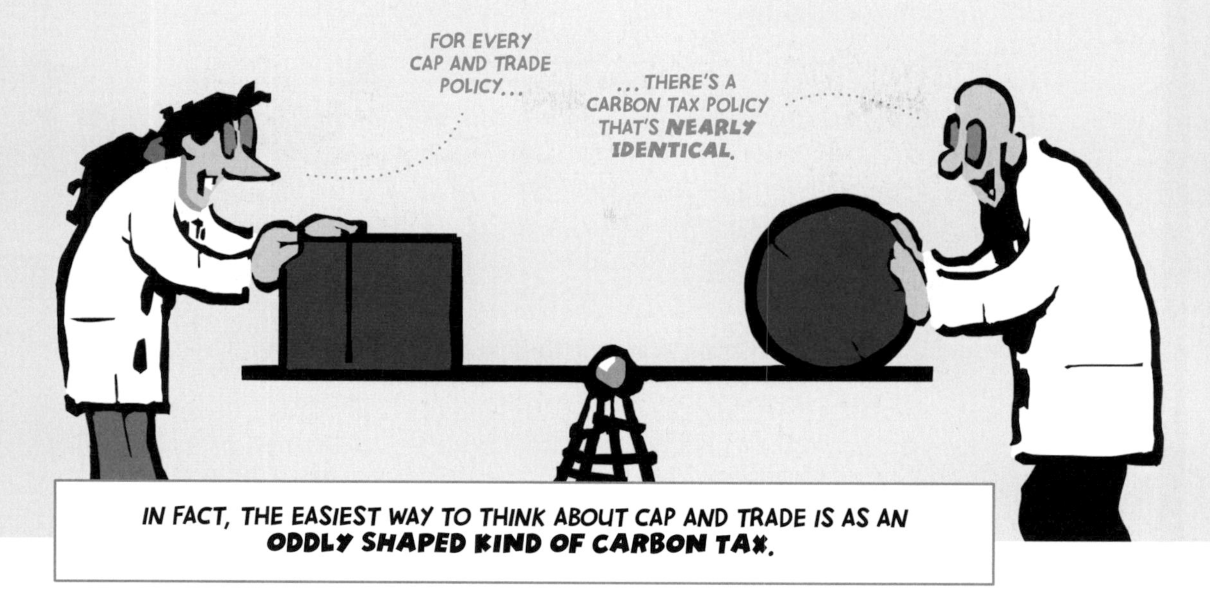
FOR EVERY CAP AND TRADE POLICY...
...THERE'S A CARBON TAX POLICY THAT'S NEARLY IDENTICAL.

IN FACT, THE EASIEST WAY TO THINK ABOUT CAP AND TRADE IS AS AN **ODDLY SHAPED KIND OF CARBON TAX.**

YOU CAN THINK OF IT AS CAP AND TAX.

TO UNDERSTAND THE **GUTS** OF CAP AND TRADE, THOUGH, A GOOD PLACE TO START IS WITH **FISH.**

TO AVOID OVERFISHING, GOVERNMENTS AROUND THE WORLD **SET LIMITS ON FISHING** AND THEN ISSUE **PERMITS.**

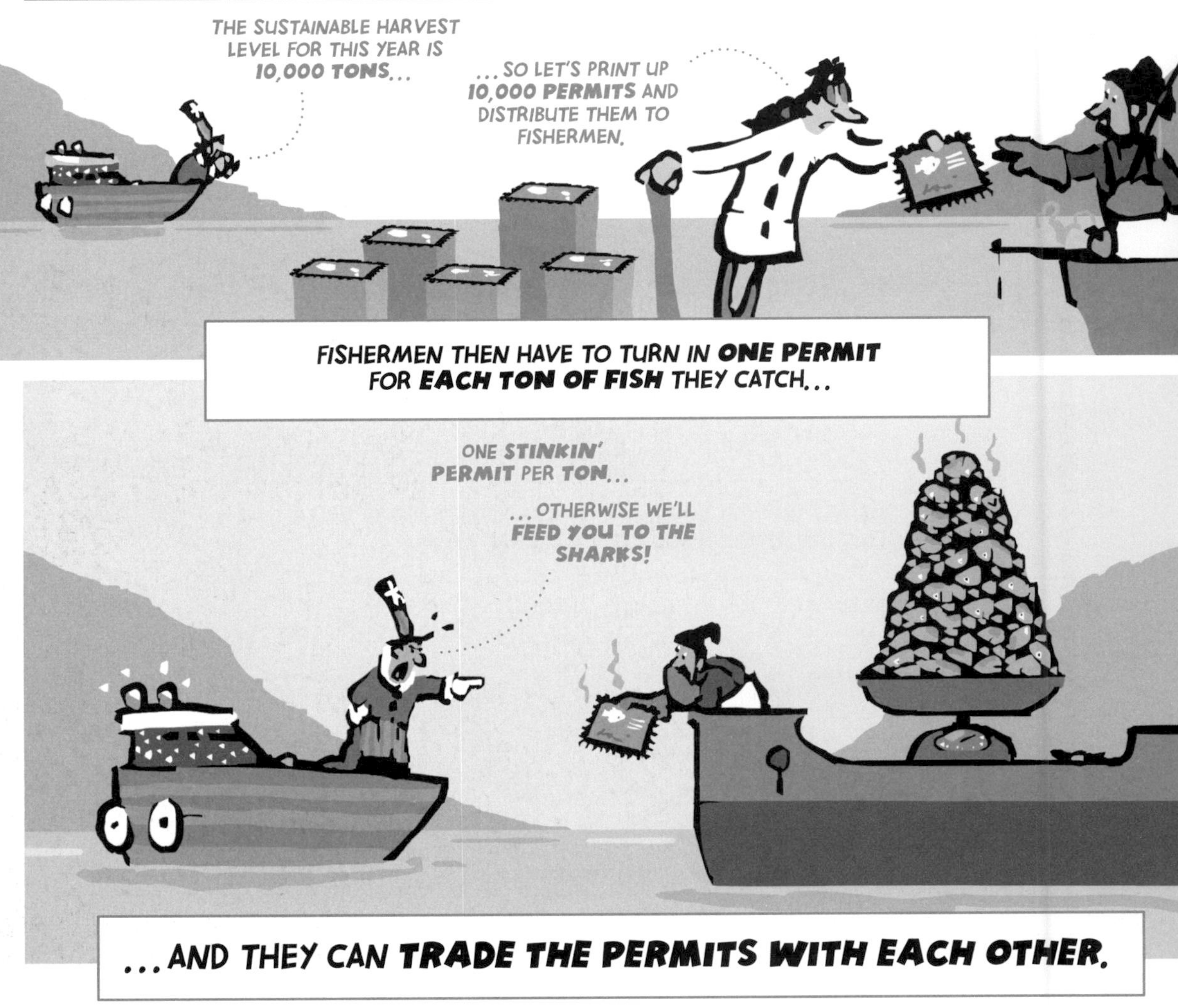
THE SUSTAINABLE HARVEST LEVEL FOR THIS YEAR IS 10,000 TONS...
...SO LET'S PRINT UP 10,000 PERMITS AND DISTRIBUTE THEM TO FISHERMEN.
FISHERMEN THEN HAVE TO TURN IN ONE PERMIT FOR EACH TON OF FISH THEY CATCH...
ONE STINKIN' PERMIT PER TON...
...OTHERWISE WE'LL FEED YOU TO THE SHARKS!
...AND THEY CAN TRADE THE PERMITS WITH EACH OTHER.

IF MY BOAT BREAKS, I CAN SELL MY PERMITS.
AND IF I FIND A TERRIFIC NEW FISHING SPOT, I CAN BUY EXTRA PERMITS.
THAT'S WHAT MAKES MARKET-BASED INSTRUMENTS FLEXIBLE!

TO SEE HOW THIS APPLIES TO CLIMATE CHANGE, **REPLACE FISH WITH CO_2.**

LET'S TRADE **CO_2...**

...JUST FOR THE **HALIBUT.**

WITH CARBON CAP AND TRADE, GOVERNMENTS **SET LIMITS ON CO_2 EMISSIONS** AND ISSUE **PERMITS.**

LET'S CAP EMISSIONS AT **100,000 TONS...**
...AND DISTRIBUTE **100,000 PERMITS TO POLLUTE.**

COMPANIES THEN HAVE TO TURN IN **ONE PERMIT** FOR **EACH TON OF CO_2** THEIR PRODUCTS EMIT...

ONE **PERMIT** PER **TON...**
...OTHERWISE WE'LL TAKE A **BITE** OUT OF YOUR **BOTTOM LINE.**

...AND THEY CAN **TRADE THE PERMITS WITH EACH OTHER.**

IF I INSTALL SOLAR PANELS, I CAN **SELL MY EXTRA PERMITS.**
AND IF I WANT TO EXPAND MY BUSINESS, I CAN **BUY EXTRA PERMITS.**

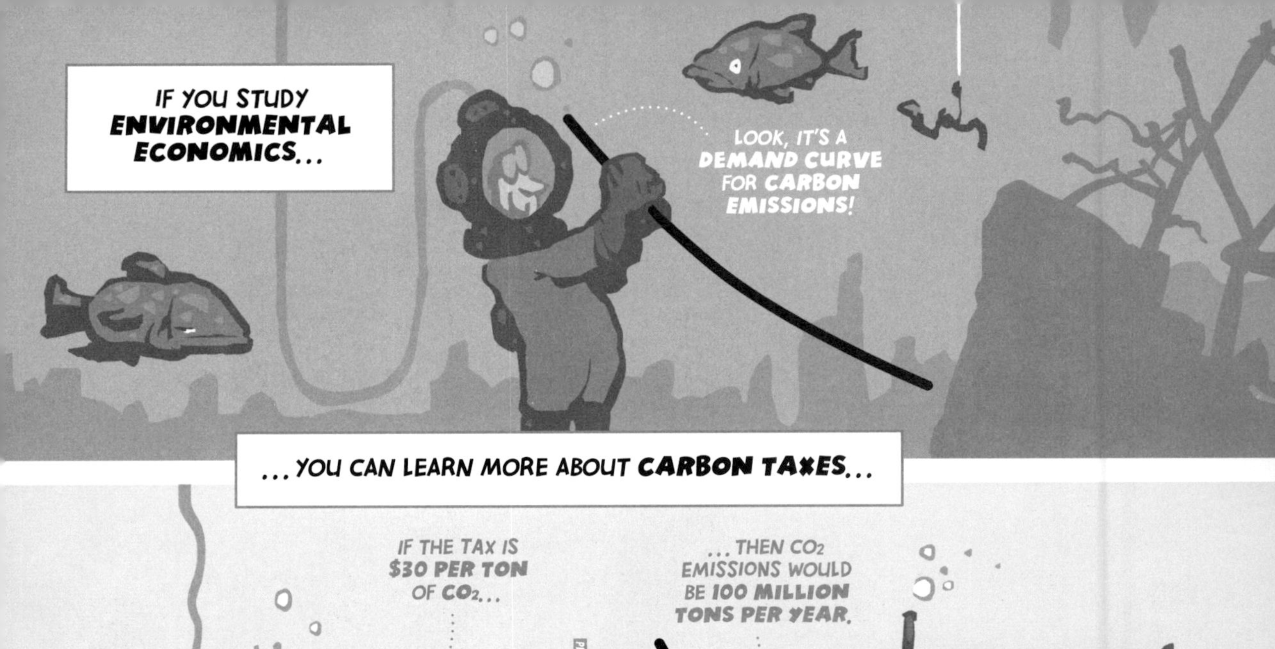
IF YOU STUDY **ENVIRONMENTAL ECONOMICS...**
LOOK, IT'S A **DEMAND CURVE** FOR **CARBON EMISSIONS!**
...YOU CAN LEARN MORE ABOUT **CARBON TAXES...**

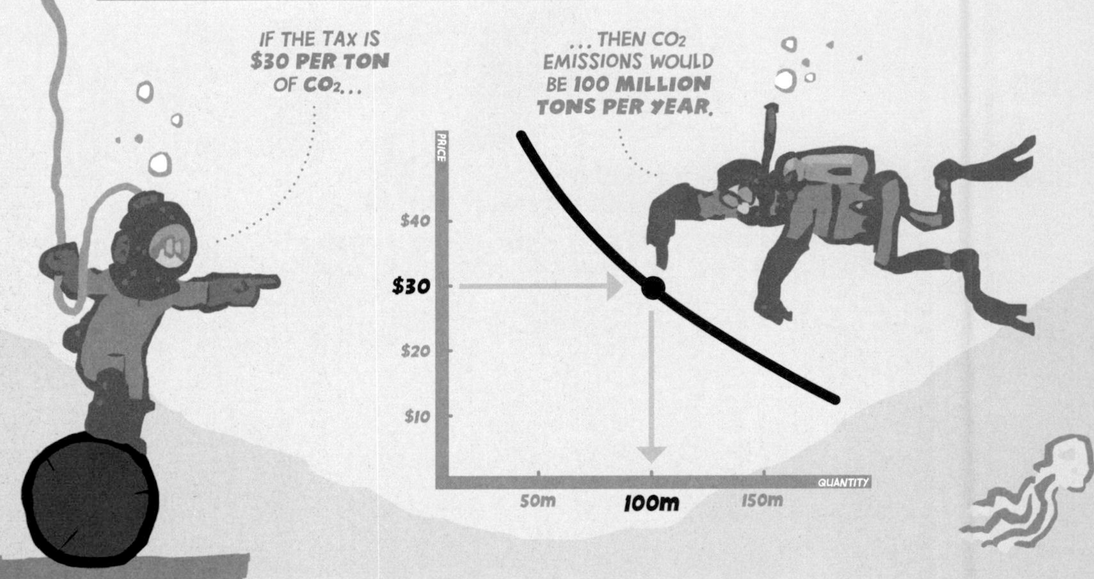
IF THE TAX IS **$30 PER TON** OF CO_2...
...THEN CO_2 EMISSIONS WOULD BE **100 MILLION TONS PER YEAR.**
PRICE
$40
$30
$20
$10
QUANTITY
50m
100m
150m

...AND HOW THEY COMPARE WITH **CAP AND TRADE SYSTEMS.**

PRICE
$40
$30
$20
$10
QUANTITY
50m
100m
150m

...IF YOU CAP CO_2 EMISSIONS AT **100 MILLION TONS PER YEAR,**

YOU'LL GET A PERMIT PRICE OF **$30 PER TON...**

BUT IF ALL THAT GIVES YOU THE **BENDS**...

...JUST REMEMBER THAT **CARBON TAXES** AND **CAP AND TRADE SYSTEMS** ARE **DIFFERENT MEANS** TO THE **SAME END.**
THEY'RE BOTH **ECONOMIC INSTRUMENTS**...
...THAT **PUT A PRICE ON CARBON.**
JUST ABOUT ANYTHING YOU CAN DO WITH **ONE**...
...YOU CAN DO WITH **THE OTHER.**

IN PARTICULAR, A **CAP AND TRADE SYSTEM** WITH **AUCTIONED PERMITS**...
...GENERATES REVENUE FOR THE GOVERNMENT, **JUST LIKE A CARBON TAX.**
AND WE CAN PUT THAT REVENUE TO GOOD USE...
...FOR EXAMPLE BY **REDUCING EXISTING TAXES.**

IT'S NO SURPRISE THAT MANY **ECONOMISTS** ARE FOND OF **ECONOMIC INSTRUMENTS**...

...BUT **DIRECT REGULATIONS** ALSO HAVE SUPPORTERS.

"RAISING THE PRICE OF CARBON IS A NECESSARY AND SUFFICIENT STEP FOR TACKLING GLOBAL WARMING."
FUND CLEAN ENERGY R&D!
PASS ENERGY EFFICIENCY LAWS NOW.
GET POWER PLANTS UNDER CONTROL!

AND WHILE THEY MIGHT DISAGREE ABOUT **POLITICS** AND **POLICY**...

WHO'S GOING TO VOTE FOR MORE REGULATION?
WHO'S GOING TO VOTE TO RAISE GAS PRICES?
YOU'VE GOT TO TRUST MARKET FORCES.
BUT WHAT ABOUT MARKET FAILURES?

...THEY SHARE A **COMMON GOAL.**

WE NEED TO DEVELOP A **LOW-CARBON ECONOMY.**

CHAPTER 15
BEYOND FOSSIL FUELS

WE NEED TO GET **BEYOND FOSSIL FUELS!**

OKAY, THERE'S **DEFORESTATION**, AND **METHANE**, AND **SF_6**, AND...

THE **MAIN HUMAN CONTRIBUTION** TO CLIMATE CHANGE COMES FROM THE **CO_2** WE EMIT BY **BURNING FOSSIL FUELS.**

NYAA!
CO_2

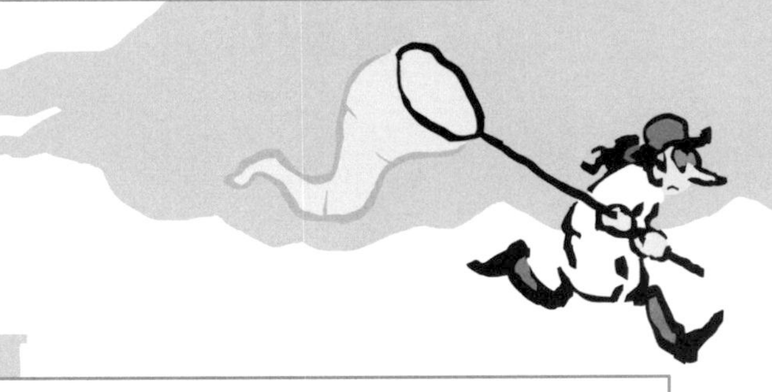

BUT THERE ARE **OTHER HUMAN ACTIVITIES** THAT **ADD CO_2** TO THE ATMOSPHERE...

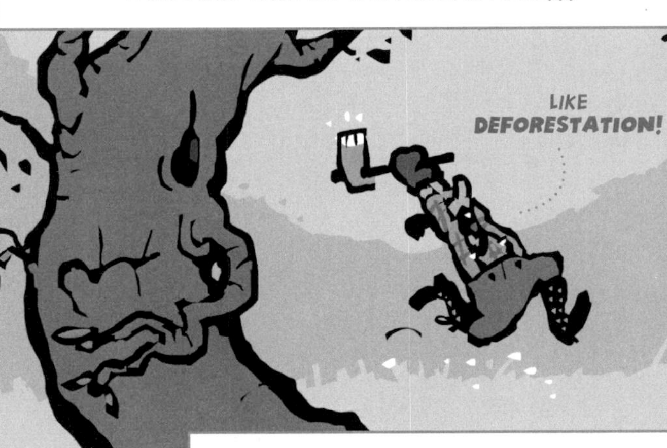
LIKE **DEFORESTATION!**

...AND THERE ARE **OTHER GREENHOUSE GASES** BESIDES **CO_2.**

MOLECULE FOR MOLECULE, I PRODUCE **28 TIMES MORE WARMING** THAN CO_2.
CH_4

YEAH, WELL I PRODUCE **24,000 TIMES MORE!**
SF_6

SO WHILE CO_2 FROM FOSSIL FUELS IS THE **KINGPIN**...

CO2
WANTED

...WE ALSO NEED TO TACKLE THE **NON–FOSSIL–FUEL SOURCES** THAT WE'LL LEARN ABOUT IN THIS CHAPTER.

EVERY LITTLE BIT **HELPS!**
AND SOMETIMES WE CAN GET MORE BANG FOR OUR BUCK BY GOING AFTER THE **LITTLE GUYS**.

DEFORESTATION INCREASES ATMOSPHERIC CO2...
CO2
CO2
CO2
...BY DISRUPTING PART OF THE NATURAL CARBON CYCLE.
FORESTS SUCK IN CARBON AND STORE IT.
WHEN WE CUT THEM DOWN, THAT CARBON FLOWS INTO THE ATMOSPHERE.

NOWADAYS, DEFORESTATION MOSTLY HAPPENS IN **POOR COUNTRIES** LIKE BRAZIL, INDONESIA, AND NIGERIA...

...BUT THEY'RE JUST FOLLOWING THE PATH **LAID DOWN BY THE RICH WORLD** IN **PREVIOUS CENTURIES.**

FORTUNATELY, MANY MIDDLE- AND HIGH-INCOME COUNTRIES ARE NOW **PLANTING NEW FORESTS...**

NEW TREES CAN STORE CARBON, JUST LIKE THE OLD ONES DID.

IT'S THE GREEN GREAT WALL OF CHINA.

...AND SOME RICH COUNTRIES ARE **PAYING POOR COUNTRIES** TO PRESERVE THEIR FORESTS.

NORWAY IS PAYING $1 BILLION...

...TO REDUCE DEFORESTATION IN INDONESIA.

METHANE (CH_4) IS THE MAIN COMPONENT OF **NATURAL GAS.**

BUT IT'S NOT JUST A FOSSIL FUEL.

IT ALSO COMES FROM FARM ANIMALS...

...AND LANDFILLS...

...AND RICE PADDIES.

WHEN WE **BURN METHANE**, IT ADDS CO_2 TO THE ATMOSPHERE.

CH4
O2
O2
H2O
H2O

CO2

BUT WHEN WE RELEASE METHANE INTO THE ATMOSPHERE **WITHOUT BURNING IT...**

...THAT'S **EVEN WORSE.**

MOLECULE FOR MOLECULE, I TRAP LOTS MORE HEAT THAN CO2.
CH4

METHANE IS A **RISK** BECAUSE IT CAN BE PART OF A **POSITIVE FEEDBACK LOOP**...

HIGHER TEMPERATURES **RELEASE METHANE** FROM THE **FROZEN ARCTIC**...
MORE ATMOSPHERIC METHANE CAUSES **MORE GLOBAL WARMING**...

WARMING TURNS INTO **EVEN MORE WARMING**.

...BUT IT'S ALSO AN **OPPORTUNITY**, ESPECIALLY AROUND **LANDFILLS**.
LET'S **CAP** THIS LANDFILL...
...AND **CAPTURE** THE METHANE THAT COMES FROM IT.

CAPTURING AND BURNING METHANE IS A **GOOD THING**...

CH4

IT'S BETTER TO **BURN FUGITIVE METHANE**...
...THAN TO **ADD IT TO THE ATMOSPHERE**.

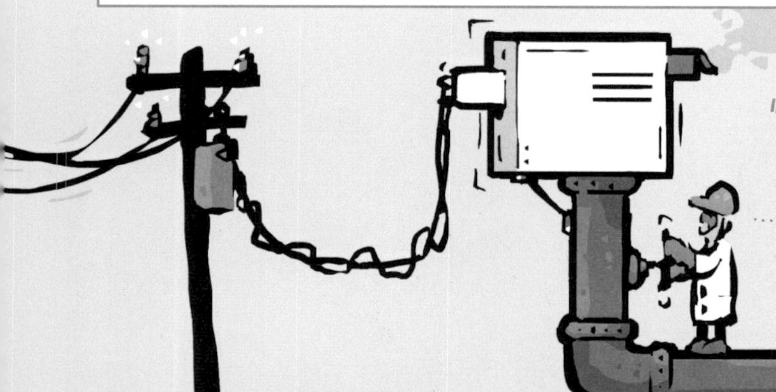
...AND IT'S **DOUBLY GOOD** IF WE CAN USE THAT METHANE TO **GENERATE ELECTRICITY**.
INSTEAD OF LETTING THE METHANE ESCAPE AND **BURNING COAL**...
...WE'RE BURNING THE METHANE AND **LEAVING THE COAL IN THE GROUND!**

MANY **OTHER SUBSTANCES** ARE ALSO INVOLVED IN CLIMATE CHANGE.

SOME OF THEM **WARM THE PLANET.**

SF_6 IS AN INDUSTRIAL GAS USED IN **ELECTRICAL EQUIPMENT...**
...AND MOLECULE FOR MOLECULE IT'S 24,000 TIMES **MORE POTENT THAN CO_2.**
SF_6

SOME OF THEM **COOL THE PLANET.**

SO_2

LIKE THE **SULFUR AEROSOLS** FROM VOLCANIC ERUPTIONS.

HENCE THE GARDEN HOSE IDEA ON PAGE 146.

SOME OF THEM ARE **SHORT LIVED.**

TWO YEARS AND I'M **OUTTA HERE!**
BLACK CARBON

SOME OF THEM ARE **LONG LIVED.**

I'M GOING TO HAUNT YOU **FOREVER.**
SF_6

THESE PETTY CRIMINALS ARE NOT AS IMPORTANT AS THE BIG THREE ...
THAT'S METHANE...
... CO2 FROM DEFORESTATION...
... AND CO2 FROM FOSSIL FUELS.
... BUT WE SHOULD STILL TRY TO GET THEM UNDER CONTROL.

RAID!
ONE WAY TO DO THAT IS WORTH DISCUSSING IN MORE DETAIL.

OFFSETS

YOU MAY HAVE ENCOUNTERED **OFFSETS** IN THE CONTEXT OF YOUR OWN PERSONAL **CARBON FOOTPRINT.**

IF I **BUY OFFSETS** THEN SOMEBODY WILL **PLANT TREES...**
... SO THAT I CAN **FLY AROUND THE WORLD...**
... AND LIVE IN A **MANSION...**
... GUILT—FREE!

BUT OFFSETS ARE ALSO USED IN MANY **CAP AND TRADE** SYSTEMS.

INSTEAD OF **BUYING EXTRA EMISSIONS PERMITS** FOR MY FACTORY...
... I'M GOING TO **BUY OFFSETS** SO SOMEBODY WILL **CAP LANDFILLS...**
... OR REDUCE CO2 EMISSIONS IN **CHINA** OR **INDIA.**

THE THEORY BEHIND OFFSETS MAKES **GOOD SENSE**...

I REALLY WANT TO GO TO HAWAII...
...SO I'D HAPPILY PAY SOMEONE $50 TO PLANT ENOUGH TREES TO OFFSET MY CO2 EMISSIONS.

CUTTING EMISSIONS IS CHEAPER IN A CHINESE FACTORY THAN IN MY FACTORY AT HOME...
...SO OFFSETS GENERATE MORE EMISSIONS REDUCTIONS PER DOLLAR.

...BUT THE **DEVIL IS IN THE DETAILS.**

HOW DO YOU KNOW THOSE TREES ARE ACTUALLY GOING TO GET PLANTED?
AND WHAT IF THEY GET PAVED OVER NEXT YEAR?

HOW DO YOU KNOW THAT CHINESE FACTORY IS REALLY GOING TO CUT ITS EMISSIONS?
AND WHAT IF IT GETS REPLACED NEXT YEAR WITH A DIRTY FACTORY ELSEWHERE?

AND WHAT ABOUT SCAMS?
HEY BUDDY, WANNA BUY SOME FOREST OFFSETS?
HEY BUDDY, WANNA BUY SOME FACTORY OFFSETS?

IN SUM, WHILE **OFFSETS** AND OTHER EFFORTS TO TACKLE NON-FOSSIL-FUEL EMISSIONS MIGHT HELP **A BIT**...

...WE HAVE TO MAKE SURE THEY **DON'T DISTRACT US FROM OUR MAIN TASK.**

CO_2

WE NEED TO **REDUCE** CO_2 FROM **FOSSIL FUELS.**

CHAPTER 16
THE CHALLENGE

IN THIS BOOK WE'VE STUDIED THE **SCIENCE** OF CLIMATE CHANGE...

...AND GLIMPSED THE **FUTURE** WE'RE LIKELY TO FACE UNDER **BUSINESS AS USUAL**...

4°C
A User's Guide to the Consequences of Business as Usual Between Now & 2100

...AND SEEN WHAT WE CAN DO TO CREATE A **DIFFERENT FUTURE**.

WE DON'T HAVE TO JUST SUFFER.
WE CAN MITIGATE AND ADAPT.

FOR OUR CONCLUSION, LET'S STEP BACK AND LOOK AT THE BIG PICTURE...
...BY THINKING ABOUT THE SIMILARITIES BETWEEN PLANET EARTH...
...AND A COMPOST HEAP.
A COMPOST HEAP?

A PROPERLY MAINTAINED **COMPOST HEAP** CONTAINS A HOST OF **MICRO-ORGANISMS.**

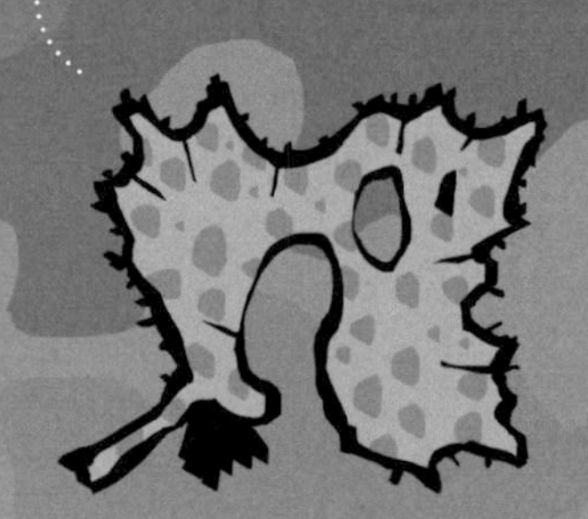
WHAT DO YOU CALL A SINGLE-CELLED ORGANISM **IN A PILE OF LEAVES?**
RUSSELL.

WHAT DO YOU CALL A SINGLE-CELLED ORGANISM **IN A CLUMP OF DIRT?**
CLAUDE.
WHAT DO YOU CALL A SINGLE-CELLED ORGANISM **WHO'S GARDENING?**
NEIL.

THOSE MICRO-ORGANISMS **BREAK DOWN LEAVES AND FOOD WASTE...**

HERE AT **OPEN-MIC NIGHT...**
...IT'S NOT JUST OLD JOKES THAT GET RECYCLED.

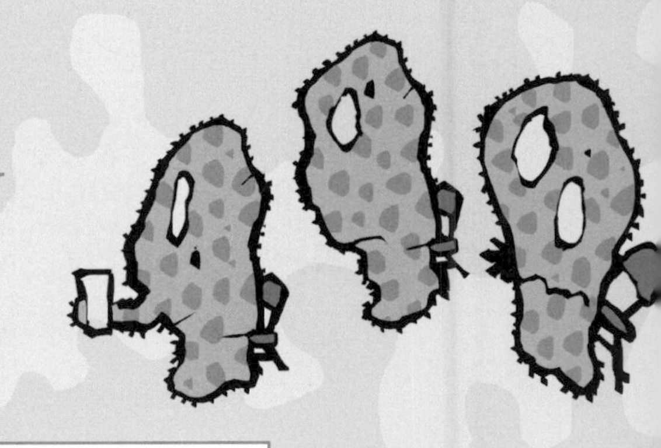

...AND TURN GARBAGE INTO GOLD.

EUREKA!
IT'S **FERTILIZER!**

BUT AS THOSE MICRO-ORGANISMS EAT...
YOUR JOKES STINK!
YUM! I LOVE ROTTEN TOMATOES!
...AND REPRODUCE...
HEY, AT LEAST I'M CRACKING MYSELF UP.
...THEY ALSO GENERATE HEAT.
PARTY ON!
IN FACT, WITHOUT PROPER MAINTENANCE, A COMPOST PILE WILL GET HOTTER AND HOTTER...
IT'S NOT RED HOT...
...IT'S INFRARED HOT!
...UNTIL THE MICRO-ORGANISMS LITERALLY KILL THEMSELVES OFF.
TOO BAD WE NEVER LEARNED ABOUT THE TRAGEDY OF THE COMMONS!

SIMILARLY, PLANET EARTH CONTAINS A HOST OF PEOPLE EATING...

WHAT'S YOUR NAME?
PATTY.
...AND REPRODUCING...

WHAT'S YOUR NAME?
RANDY.
...AND GENERATING HEAT.

WHAT'S YOUR NAME?
PHIL.
JUST LIKE MICRO-ORGANISMS WARMING A COMPOST PILE, WE'RE WARMING THE PLANET.

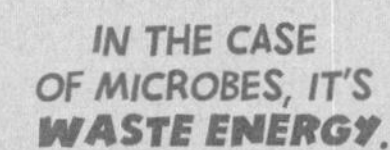
IN THE CASE OF MICROBES, IT'S WASTE ENERGY.

IN THE CASE OF HUMANS, IT'S EMISSIONS OF GREENHOUSE GASES.

FORTUNATELY, **UNLIKE** THE MICRO-ORGANISMS, WE'VE GOT SOME **SECRET WEAPONS.**

IF WE ONLY HAD **BRAINS!**

WE HAVE SMART SCIENTISTS INVENTING **NEW TECHNOLOGIES.**

BIOFUELS

CARBON CAPTURE

NEXT-GENERATION RENEWABLES

SUPER-EFFICIENT VEHICLES

WE HAVE GOOD **POLICY IDEAS** TO GUIDE OUR WAY.

CARBON PRICING!
Carbon Tax
Cap and Trade

AND WE HAVE THE ABILITY TO **LOOK AHEAD** AND MAKE **SMART CHOICES ABOUT THE FUTURE.**

BUSINESS AS USUAL ONLY GOT **ONE STAR** IN OUR GUIDEBOOK.
Business as usual
Alternatives

MAKE NO MISTAKE, THOUGH: OUR TASK IS **DAUNTING.**

HAVING **SECRET WEAPONS** DOESN'T GUARANTEE YOU'RE GOING TO **WIN.**

AS WITH THE MICRO-ORGANISMS, THE AMOUNT OF WARMING CAUSED BY HUMANS IS RELATED TO **POPULATION...**

THE HUMAN POPULATION IS LIKELY TO INCREASE BY OVER **50%** THIS CENTURY.

...AND TO **ACTIVITY LEVELS.**

ECONOMIC GROWTH IS GOING TO BRING A LOT OF PEOPLE **OUT OF POVERTY.**
THAT'S GOING TO BE **TERRIFIC...**
...BUT IT WILL ALSO MEAN A LOT MORE **ENERGY USE.**

THAT'S WHY BUSINESS AS USUAL COULD RESULT IN A **TRIPLING** OF ANNUAL CO_2 EMISSIONS THIS CENTURY.

IT WILL TAKE **TREMENDOUS EFFORT** JUST TO KEEP ANNUAL EMISSIONS STEADY.

AND REMEMBER THAT CO_2 STAYS IN THE ATMOSPHERE FOR A **LONG, LONG TIME...**

SO EVEN IF WE DO KEEP CO2 **EMISSIONS STEADY...**
...**ATMOSPHERIC CONCENTRATIONS** WILL STILL RISE.

...AND THAT THE PROBLEM IS COMPOUNDED BY **DEFORESTATION...**

...AND **FEEDBACKS...**

...AND THAT ANY SOLUTIONS WILL HAVE TO **OVERCOME THE TRAGEDY OF THE COMMONS.**

THEY WEREN'T KIDDING: THE TASK **IS DAUNTING.**

BUT **DAUNTING** DOESN'T MEAN **IMPOSSIBLE!**

OK, BUT WHAT CAN **WE** DO?

YOU CAN HELP BY **DREAMING OF A BETTER FUTURE...**

I SEE A PRINCESS.
I SEE A DRAGON.
I SEE A SOLUTION TO CLIMATE CHANGE!

...AND THEN FIGURING OUT WHETHER THAT **DREAM** COULD REALLY **COME TRUE...**

THIS WAS MORE FUN WHEN WE WERE JUST DREAMING.

...AND THEN WORKING TO **MAKE IT HAPPEN.**

OKAY, IT'S **READY TO GO.**

TIME TO **CHANGE THE WORLD!**

AS YOU KNOW, LOTS OF ECONOMISTS DREAM OF **CARBON PRICING**...

IF THE RICH WORLD ADOPTS **REVENUE-NEUTRAL CARBON TAXES**...
...THEN THE **POWER OF CAPITALISM** COULD BRING US...
...**CHEAP, CLEAN ENERGY** TO POWER THE WHOLE WORLD.
CHECKMATE!

...BUT IT'S OKAY IF YOUR DREAM IS **DIFFERENT**.

ALL-OUT MOBILIZATION
WE NEED A **WAR** ON CARBON EMISSIONS.

GOVERNMENT-FUNDED CLEAN ENERGY RESEARCH

WE NEED A **BREAKTHROUGH**.

PLANNING AND ADAPTATION
WE NEED TO **PREPARE**.

LOW-CARBON LIFESTYLES
WE NEED TO **LIVE SIMPLY** SO THAT OTHERS CAN **SIMPLY LIVE**.

YOU CAN **ALSO** HELP BY LOOKING AT **YOUR OWN LIFE**...

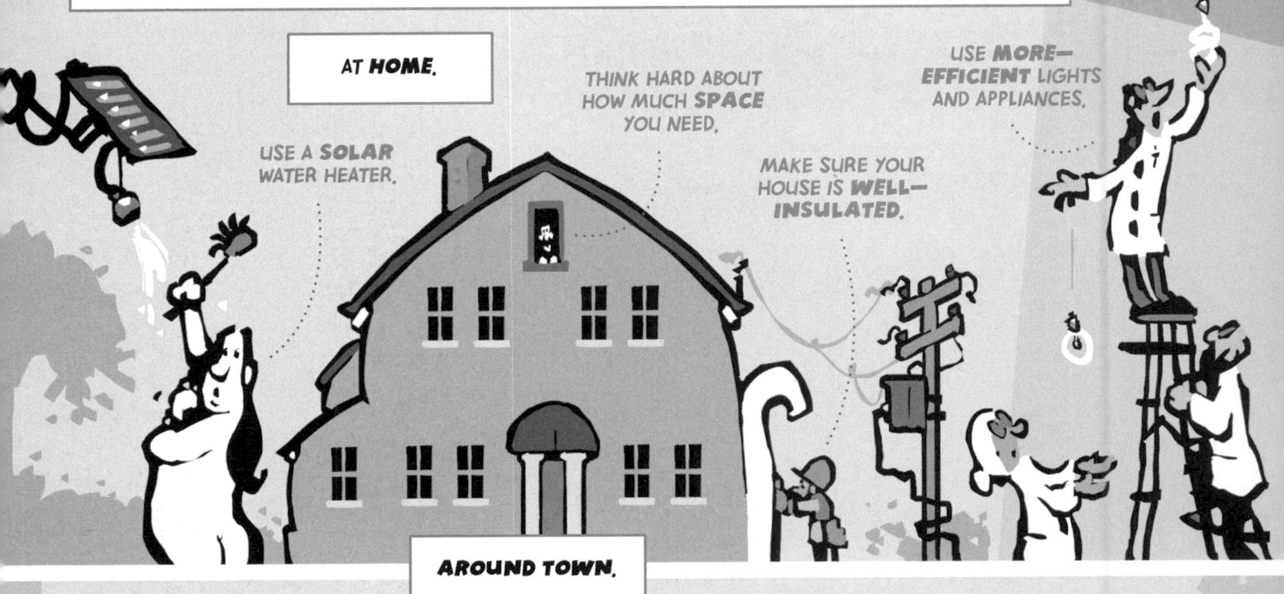
AT **HOME**.
USE A **SOLAR** WATER HEATER.
THINK HARD ABOUT HOW MUCH **SPACE** YOU NEED.
MAKE SURE YOUR HOUSE IS **WELL-INSULATED**.
USE **MORE-EFFICIENT** LIGHTS AND APPLIANCES.

AROUND TOWN.
REDUCE, REUSE, RECYCLE.
THINK HARD ABOUT HOW MUCH **STUFF** YOU NEED TO BUY.
EAT **LESS** MEAT.
EAT **MORE** VEGETABLES.

ON THE ROAD.

DON'T **SPEED**...
...AND KEEP YOUR TIRES **INFLATED**.
STOP

THINK HARD ABOUT HOW MUCH YOU **FLY**.

DRIVE **LESS**.
HAVE **MORE** LOW-CARBON ADVENTURES.

...AND BY HELPING YOUR COMMUNITY AND YOUR COUNTRY **MAKE SMART DECISIONS.**

USE YOUR **VOICE.**

EXCUSE ME, **REPRESENTATIVE...**

JOIN A **CHORUS.**

Energy & Enterprise Initiative

Sierra Club
Citizens Climate Lobby

INVITE OTHERS TO **JOIN YOU.**

WHEN YOU'RE DONE WITH THIS BOOK, GIVE IT TO A **FRIEND** OR A **NEIGHBOR...**
...OR AN **ELECTED OFFICIAL.**

IN WAYS BIG AND SMALL, **YOU** CAN HELP WRITE THE NEXT CHAPTER.

THAT NEXT CHAPTER IS BEING WRITTEN **NOW**.

GOOD LUCK TO US **ALL**.

GLOSSARY

A

ADAPTATION
EFFORTS TO REDUCE THE IMPACT OF CLIMATE CHANGE, SUCH AS BUILDING SEAWALLS OR DESIGNING DROUGHT-RESISTANT FARMING PRACTICES; CONTRAST WITH **MITIGATION**: 102, 105–10, 184

AIR
MEASURED IN TERMS OF THE PERCENTAGES OF DIFFERENT TYPES OF MOLECULES, THE ATMOSPHERE IS ABOUT 78% NITROGEN (N_2), 21% OXYGEN (O_2), AND 1% OTHER, INCLUDING ABOUT 0.04% CARBON DIOXIDE (CO_2), WHICH EQUALS 400 PARTS PER MILLION (ppm) CO_2: 40

ALBEDO
THE REFLECTIVITY OF A SUBSTANCE; THE EARTH'S ALBEDO IS ABOUT 30%. TO REMEMBER ITS MEANING, NOTE THAT "ALBEDO" ("WHITENESS") HAS THE SAME LATIN ROOT AS "ALBINO": 56, 146

THE **ICE-ALBEDO EFFECT** IS A POSITIVE FEEDBACK LOOP THAT CONTRIBUTES TO THE COMINGS AND GOINGS OF ICE AGES: IN THE GLACIAL DIRECTION, ACCUMULATING ICE AND SNOW INCREASES THE ALBEDO OF THE EARTH'S SURFACE, RESULTING IN DECREASED SOLAR ABSORPTION THAT LOWERS TEMPERATURES AND LEADS TO MORE ACCUMULATION OF ICE AND SNOW; IN THE INTERGLACIAL DIRECTION, MELTING ICE AND SNOW REDUCES THE ALBEDO OF THE EARTH'S SURFACE, RESULTING IN INCREASED SOLAR ABSORPTION THAT RAISES TEMPERATURES AND LEADS TO MORE MELTING ICE AND SNOW: 35, 37, 126

ANTHROPOGENIC
CAUSED BY HUMANS: 70–72

ARCTIC
THE REGION AROUND THE NORTH POLE; MOST **ARCTIC ICE** IS FLOATING, AND SO MELTING DOES NOT CONTRIBUTE TO SEA LEVEL RISE (UNLIKE THE MOSTLY LAND-BASED **ANTARCTIC ICE** AROUND THE SOUTH POLE): 86, 94, 124

ARRHENIUS, SVANTE
A SWEDISH CHEMIST, SVANTE ARRHENIUS (1859–1927) WAS ONE OF THE FIRST SCIENTISTS TO STUDY ATMOSPHERIC CO_2: 62, 119

B

BLACK CARBON
AIRBORNE PARTICLES SUCH AS SOOT, MOST COMMONLY FORMED FROM BURNING PLANT MATTER OR FOSSIL FUELS: 178

BORDER TAX ADJUSTMENT
A TAX THAT COULD BE APPLIED (AT LEAST IN THEORY) TO THE CARBON "EMBODIED" IN IMPORTED GOODS; FOR EXAMPLE, IF COUNTRY A HAS A CARBON TAX BUT COUNTRY B DOESN'T, COUNTRY A COULD TAX GOODS IMPORTED FROM COUNTRY B: 144

BUSINESS AS USUAL
A HYPOTHETICAL FUTURE FEATURING MINIMAL EFFORTS TO REDUCE CARBON EMISSIONS AND STRONG ECONOMIC GROWTH BASED ON FOSSIL FUELS; FOR EXAMPLE, SEE THE IPCC SCENARIO KNOWN AS RCP 8.5: 79–82, 128–29, 189

C

CAP AND TRADE
SEE **CARBON PRICING**

CARBON (C)
ONE OF THE BUILDING BLOCKS OF LIFE ON EARTH. BURNING PLANTS AND FOSSIL FUELS TURNS THE CARBON THEY CONTAIN INTO CARBON DIOXIDE (CO_2), WHICH IS A GREENHOUSE GAS. THE **CARBON CYCLE** DESCRIBES HOW CARBON CIRCULATES AROUND THE PLANET AND INCLUDES NOT ONLY BURNING OF PLANTS AND FOSSIL FUELS BUT ALSO PROCESSES SUCH AS PHOTOSYNTHESIS; AIR-OCEAN GAS EXCHANGE, WHICH IS RELATED TO OCEAN ACIDIFICATION; THE CREATION AND WEATHERING OF ROCKS, ETC.

CARBON CYCLE: 18, 42–43, 47
CARBON DIOXIDE (CO_2): 17, 21, 39–50, 114, 191
CARBON SINK: 46, 146
CARBON VERSUS CO_2: 44 (SEE ALSO **CARBON VERSUS CO_2**)

CARBON PRICING
A GOVERNMENT POLICY THAT SEEKS TO REDUCE CARBON EMISSIONS BY MAKING POLLUTING EXPENSIVE: 143–44, 159–70, 189, 193

THE SIMPLEST FORM OF CARBON PRICING IS A **CARBON TAX** ON THE CARBON CONTENT OF FOSSIL FUELS: 162–63

THE OTHER FORM OF CARBON PRICING IS **CAP AND TRADE**, A POLICY THAT REQUIRES POLLUTERS TO TURN IN ONE PERMIT FOR EACH TON OF CO_2 THEY EMIT; THE NUMBER OF PERMITS IS LIMITED (RAISING OR LOWERING THIS "CAP" MAKES THE POLICY LESS OR MORE STRICT) AND PERMITS CAN BE TRADED BASED ON SUPPLY AND DEMAND: 164–67, 180–81

CAUTIONARY NOTE: A CARBON PRICE OF (SAY) $30 PER TON OF **CARBON** IS VERY DIFFERENT FROM A CARBON PRICE OF $30 PER TON OF **CARBON DIOXIDE** (SEE BELOW) SO IT'S IMPORTANT TO CLEARLY IDENTIFY WHICH OF THESE IS BEING DISCUSSED. PRICES ARE MOST COMMONLY GIVEN PER TON OF CARBON DIOXIDE, SO AN UNCLEAR REFERENCE TO "A TAX OF $30 PER TON" PROBABLY MEANS $30 PER TON OF CARBON DIOXIDE.

CARBON VERSUS CO_2

A COMMON SOURCE OF CONFUSION. CARBON (C) CONTAINS 6 PROTONS AND 6 NEUTRONS, FOR AN ATOMIC MASS OF 12. THE ATOMIC MASS OF OXYGEN IS 16, SO CARBON DIOXIDE (CO_2) HAS A MOLECULAR MASS OF 44. PUT THESE TOGETHER TO SEE THAT X TONS OF CO_2 CONTAINS ONLY $12X/44 \approx 0.27X$ TONS OF CARBON. SIMILARLY, BURNING X TONS OF CARBON CREATES $44X/12 \approx 3.67X$ TONS OF CO_2. AS A RESULT, A CARBON PRICE OF \$X PER TON OF CARBON EQUALS ONLY $\$12X/44 \approx \$0.27X$ PER TON OF CO_2. (CARBON PRICES ARE USUALLY STATED PER TON OF CO_2, AND IN MOST CASES THE TONS ARE METRIC TONS, NOT SHORT TONS.)

CO_2

CARBONIFEROUS PERIOD

THE TIME PERIOD FROM 360–300 MILLION YEARS AGO: **23**

CATCH-UP

THE IDEA—ALSO CALLED **CONVERGENCE**—THAT POOR COUNTRIES WILL HAVE FASTER ECONOMIC GROWTH THAN RICH COUNTRIES, CAUSING THEIR ECONOMIES TO CATCH UP: **4–6, 81, 110**

CFCs

PREVIOUSLY USED IN AEROSOL CONTAINERS AND AS REFRIGERANTS, CHLOROFLUOROCARBONS (CFCs) DAMAGE THE OZONE LAYER AND ARE BEING PHASED OUT: **21**

CH_4 (METHANE)

SEE **GREENHOUSE GAS**

CLEAN ENERGY

IN THE CONTEXT OF CLIMATE CHANGE, CLEAN ENERGY IS ENERGY THAT DOESN'T (OR DOESN'T MUCH) INCREASE ATMOSPHERIC CO_2 CONCENTRATIONS; THIS INCLUDES RENEWABLE ENERGIES SUCH AS HYDROPOWER, WIND, AND SOLAR, PLUS NUCLEAR POWER, CCS (CARBON CAPTURE AND STORAGE), AND ENERGY EFFICIENCY: **150**

CLIMATE

AVERAGE WEATHER, TYPICALLY OVER 30-YEAR PERIODS, INCLUDING AVERAGE VALUES AND EXPECTED VARIABILITY FOR DAILY HIGH AND LOW TEMPERATURES, PRECIPITATION, ETC.: **12**

CLIMATE CHANGE

A CHANGE IN GLOBAL AVERAGE TEMPERATURE AND OTHER CHANGES IN THE EARTH SYSTEM, EITHER AS A RESULT OF NATURAL PROCESSES OR (AS WITH **ANTHROPOGENIC CLIMATE CHANGE**) AS A RESULT OF BURNING FOSSIL FUELS AND OTHER HUMAN ACTIVITIES: **12**

CLIMATE MODELS

COMPUTER MODELS OF THE EARTH'S CLIMATE SYSTEM: **74**

D

DEFORESTATION
CUTTING DOWN TREES FOR FARMING, URBAN DEVELOPMENT, OR OTHER HUMAN ACTIVITIES: 10, 46, 172, 174–75, 191

DEVELOPED/DEVELOPING COUNTRIES
THE DIVIDING LINE CAN BE BLURRY (SEE **CATCH–UP**), BUT THE DEVELOPED WORLD INCLUDES NORTH AMERICA, EUROPE, JAPAN, AND OTHER RICH COUNTRIES; TOGETHER THEY ACCOUNT FOR ONE–FIFTH OF THE WORLD'S POPULATION. THE DEVELOPING WORLD INCLUDES MUCH OF ASIA, AFRICA, AND SOUTH AMERICA: 4–6, 79, 107–110, 175, 190

E

EL NIÑO
A WEATHER PATTERN IN THE TROPICAL PACIFIC OCEAN THAT AFFECTS GLOBAL TEMPERATURE AND PRECIPITATION: 72

ELECTROMAGNETIC RADIATION
A FORM OF ENERGY THAT INCLUDES VISIBLE LIGHT, LONGER–WAVELENGTH INFRARED RADIATION, AND SHORTER–WAVELENGTH ULTRAVIOLET RADIATION: 55–57

INFRARED (IR): 55, 57, 187
ULTRAVIOLET (UV): 20, 55
VISIBLE: 55

ENERGY IN/ENERGY OUT
THE CRUCIAL DETERMINANTS OF ENERGY BALANCE FOR THE PLANET. FOR EXAMPLE, IF ENERGY IN IS LARGER THAN ENERGY OUT THEN THE PLANET WILL WARM UP: 51–59

EVOLUTION
THE PROCESS BY WHICH MUTATION AND NATURAL SELECTION LEADS TO THE CREATION OF NEW SPECIES: 24, 104–105

EXTERNAL COST
COSTS IMPOSED ON A THIRD PARTY, FOR EXAMPLE IF PERSON A SELLS COAL–FIRED ELECTRICITY TO PERSON B BUT BURNING THE COAL HARMS PERSON C'S HEALTH OR PROPERTY: 155

F

FEEDBACK
A SECONDARY EFFECT THAT EITHER AMPLIFIES OR WEAKENS SOME INITIAL CHANGE. A **POSITIVE FEEDBACK LOOP** AMPLIFIES THE INITIAL CHANGE, SO THAT WARMING TURNS INTO EVEN MORE WARMING OR COOLING TURNS INTO EVEN MORE COOLING. A **NEGATIVE FEEDBACK LOOP** WEAKENS THE INITIAL CHANGE, SO THAT WARMING TURNS INTO A BIT LESS WARMING AND COOLING TURNS INTO A BIT LESS COOLING: 32–33, 35, 37, 96, 126, 177, 191

ICE–ALBEDO FEEDBACK: 35, 37, 126
WATER VAPOR FEEDBACK: 96, 126

FINGERPRINTS
TELLTALE SIGNS CONNECTING HUMAN EMISSIONS OF GREENHOUSE GASES TO CLIMATE CHANGE: 70–71

FISHERIES
A COMMON EXAMPLE OF THE **TRAGEDY OF THE COMMONS** AND OF THE USE OF CAP AND TRADE SYSTEMS TO ADDRESS SUCH PROBLEMS: **140, 143, 166**

FOSSIL FUEL
COAL, OIL, OR NATURAL GAS PRODUCED BY NATURAL PROCESSES OVER MILLIONS OF YEARS: **10, 23, 46, 80–82, 136, 148–56**

FRACKING
A NEW METHOD OF EXTRACTING FOSSIL FUELS, SHORT FOR "HYDRAULIC FRACTURING": **46**

G

GLACIAL PERIOD
SCIENTIFIC NAME FOR WHAT IS COMMONLY CALLED AN **ICE AGE**: A PERIOD OF TIME WITH EXTENSIVE GLACIERS: **25–38**

GLOBAL WARMING
SEE **CLIMATE CHANGE**

GREENHOUSE GAS
A GAS THAT DOESN'T INTERACT MUCH WITH RELATIVELY SHORT WAVELENGTH RADIATION (THE ULTRAVIOLET AND VISIBLE RADIATION INCOMING FROM THE SUN) BUT INTERACTS STRONGLY WITH RELATIVELY LONG WAVELENGTH RADIATION (THE INFRARED RADIATION OUTGOING FROM THE EARTH). THE PRINCIPAL GREENHOUSE GASES ON EARTH ARE WATER VAPOR, CARBON DIOXIDE (CO_2), AND METHANE; LESS-IMPORTANT GREENHOUSE GASES INCLUDE NITROUS OXIDE, SF_6, AND MANY OTHERS: **21, 58–62, 172–78**

METHANE (CH_4): **58, 124, 172, 176–77**
SULFUR HEXAFLOURIDE (SF_6): **172, 178**
WATER VAPOR: **40, 58, 96**

I

ICE AGES
A PERIOD OF TIME WITH EXTENSIVE GLACIERS, FOR EXAMPLE FROM ABOUT 120,000 TO 12,000 YEARS AGO; SCIENTISTS SOMETIMES REFER TO THESE AS **GLACIAL PERIODS**: **25–38**

ICE-ALBEDO EFFECT
SEE **ALBEDO**

ICE CORE
A VERTICAL "ROD" OF ICE, DRILLED FROM A GLACIER OR ICE SHEET, THAT CAN BE USED TO STUDY THE HISTORY OF THE EARTH'S CLIMATE BY EXAMINING THE CONTENTS OF DIFFERENT LAYERS THAT CORRESPOND TO DIFFERENT TIMES; FOR EXAMPLE, THE ANALYSIS OF AIR BUBBLES TRAPPED IN THE ICE CAN BE USED TO ESTIMATE CO_2 CONCENTRATIONS, AND THE PREVALENCE OF CERTAIN RARE MOLECULES (SUCH AS DEUTERIUM, A FORM OF H_2O ALSO CALLED HEAVY WATER) CAN BE USED TO ESTIMATE THINGS LIKE TEMPERATURE AND THE RATIO OF WATER TO ICE ON THE PLANET: **29**

TEMPERATURE AND CO_2 DATA: **48–50**

ICE SHEET
A LARGE GLACIER SUCH AS THE GREENLAND ICE SHEET OR THE WEST ANTARCTIC ICE SHEET: **28, 94, 117**

INFRARED (IR)
SEE **ELECTROMAGNETIC RADIATION**

INSURANCE
A METHOD OF ADDRESSING LOW-PROBABILITY RISKS (FOR EXAMPLE THE RISK OF GETTING IN A CAR ACCIDENT) BY MAKING REGULAR PAYMENTS IN EXCHANGE FOR COVERAGE IN CASE OF DISASTER; SPENDING MONEY NOW TO REDUCE CARBON EMISSIONS CAN BE THOUGHT OF AS AN "INSURANCE POLICY" AGAINST THE RISK OF A CLIMATE CHANGE DISASTER: 130–32, 137

INTERGLACIAL PERIOD
A RELATIVELY WARM PERIOD IN BETWEEN ICE AGES; THE CURRENT INTERGLACIAL PERIOD, CALLED THE HOLOCENE, STARTED ABOUT 12,000 YEARS AGO: 29, 36–38

IPCC
THE INTERGOVERNMENTAL PANEL ON CLIMATE CHANGE, AN INTERNATIONAL BODY THAT HAS STUDIED CLIMATE CHANGE SINCE 1988; THEIR FIFTH ASSESSMENT REPORT (AR5) CAME OUT IN 2014: 68, 71, 123–24

K

KEELING CURVE
MEASUREMENTS OF ATMOSPHERIC CO_2 CONCENTRATIONS, STARTED BY CHARLES DAVID KEELING IN 1958: 41–47

M

METHANE (CH_4)
SEE **GREENHOUSE GAS**

MILANKOVITCH CYCLES
VARIATIONS IN THE EARTH'S ORBIT AROUND THE SUN CAUSED BY GRAVITATIONAL INTERACTIONS WITH OTHER PLANETS; THESE VARIATIONS DON'T SIGNIFICANTLY CHANGE THE AMOUNT OF SOLAR RADIATION THAT REACHES THE EARTH OVER THE COURSE OF A YEAR, BUT THEY DO AFFECT TIMING, I.E., HOW MUCH RADIATION ARRIVES AT DIFFERENT TIMES OF THE YEAR: 30–32, 50

MILUTIN MILANKOVITCH (1879–1958) WAS A SERBIAN MATHEMATICIAN WHO ARGUED THAT THESE VARIATIONS CAUSED THE ICE AGES. OUR SIMPLIFIED VERSION OF HIS THEORY FOCUSES ON THE EFFECTS OF "STRONG SEASONS" AND "WEAK SEASONS," BUT HIS ACTUAL FOCUS WAS ON SUMMER IN THE NORTHERN HEMISPHERE (NOTABLY AT 65 DEGREES NORTH LATITUDE), WITH COLD SUMMERS IN THE NORTH LEADING TO GLACIAL ADVANCE AND WARM SUMMERS IN THE NORTH LEADING TO GLACIAL RETREAT.

MITIGATION
EFFORTS TO REDUCE ATMOSPHERIC CONCENTRATIONS OF GREENHOUSE GASES; CONTRAST WITH **ADAPTATION**: 132, 184

MONTREAL PROTOCOL
A GLOBAL TREATY SIGNED IN 1987 TO PROTECT THE OZONE LAYER BY PHASING OUT OZONE-DEPLETING CHEMICALS SUCH AS CFCS: 21, 143

N

NEGATIVE FEEDBACK LOOP
SEE **FEEDBACK**

O

OCEAN ACIDIFICATION
AN INCREASE IN THE CONCENTRATION OF HYDROGEN IONS IN THE OCEANS, SOMETIMES CALLED GLOBAL WARMING'S EQUALLY EVIL TWIN: 46–47, 98–100, 106, 147, 155

OFFSETS
THE IDEA THAT AN INDIVIDUAL, COMPANY, OR COUNTRY CAN PAY FOR CARBON REDUCTIONS ELSEWHERE (FOR EXAMPLE, BY PLANTING TREES) INSTEAD OF REDUCING THEIR OWN EMISSIONS: 179–82

OZONE (O_3)
A MOLECULE CONSISTING OF THREE OXYGEN ATOMS; OZONE EXISTS BOTH AT GROUND LEVEL, WHERE IT IS A POLLUTANT RELATED TO SMOG, AND HIGH UP IN THE ATMOSPHERE, WHERE IT FORMS THE **OZONE LAYER** THAT BLOCKS HARMFUL UV RADIATION: 19–21

P

PHENOLOGY
THE STUDY OF THE TIMING OF BIRD MIGRATIONS, PLANT FLOWERINGS, AND OTHER LIFE CYCLE EVENTS IN PLANTS AND ANIMALS: 12, 73

PHOTOSYNTHESIS
A CHEMICAL REACTION THAT USES SUNLIGHT TO SUSTAIN LIFE, MOST COMMONLY **OXYGENIC PHOTOSYNTHESIS** THAT CONSUMES CARBON DIOXIDE AND PRODUCES OXYGEN AS A BYPRODUCT, THEREBY CONTRIBUTING TO THE CARBON CYCLE: 17–18, 42–43

POPULATION
THE NUMBER OF PEOPLE ON THE PLANET, EXPECTED TO RISE TO 10 OR 11 BILLION BY 2100 FROM 6 BILLION IN 2000 AND 1.6 BILLION IN 1900: 5, 81, 118, 190

AND "FIVE CHINAS": 80–81

POSITIVE FEEDBACK LOOP
SEE **FEEDBACK**

S

SCIENTIFIC METHOD
A PROCESS OF LEARNING AND DISCOVERY THAT INVOLVES DEVELOPING AND REFINING HYPOTHESES AND TESTING THOSE HYPOTHESES AGAINST REAL-WORLD DATA: 63–74, 124

SEA LEVEL RISE
ONE OF THE RISKS ASSOCIATED WITH CLIMATE CHANGE, RISING SEA LEVELS ARE RELATED TO BOTH THE THERMAL EXPANSION OF WARMING OCEANS AND THE MELTING OF LAND-BASED ICE: 92–94, 116–17, 124

SMITH, ADAM

AUTHOR OF *THE WEALTH OF NATIONS*, ADAM SMITH (1723–1790) WAS A SCOTTISH PHILOSOPHER AND ECONOMIST WHO COINED THE "INVISIBLE HAND" METAPHOR: **4, 142**

SNOWBALL EARTH

A PLANET WITH TEMPERATURES COLD ENOUGH TO FREEZE SEAWATER, EVEN AT THE EQUATOR; SCIENTISTS DEBATE ABOUT WHETHER THE EARTH EVER GOT THIS COLD: **22, 24**

SUBSIDY

PAYMENTS TO THE PRODUCERS OF A PRODUCT; CONTRAST WITH TAXES, ESPECIALLY CARBON TAXES: **154**

SULFUR DIOXIDE (SO_2)

A GAS THAT, AFTER BEING RELEASED INTO THE ATMOSPHERE, CAN FORM SULFUR PARTICLES THAT REFLECT SOLAR RADIATION; ALSO IMPORTANT IN ACID RAIN: **146–47, 178**

T

TAX SWAP

THE IDEA OF INCREASING TAXES ON "BADS" SUCH AS POLLUTION AND USING THE TAX REVENUE TO REDUCE TAXES ON "GOODS" SUCH AS INCOME: **163, 169**

TEMPERATURE

AVERAGING SURFACE TEMPERATURES AROUND THE PLANET OVER THE COURSE OF A YEAR YIELDS THE GLOBAL AVERAGE TEMPERATURE, CURRENTLY AROUND 15°C (59°F): **52, 124**

CAUTIONARY NOTE FOR AMERICANS:

THE FAHRENHEIT SCALE (°F) USED IN THE USA DIFFERS FROM THE CELSIUS SCALE (°C) COMMONLY USED IN SCIENCE, SO EXTREME CAUTION IS NEEDED. WATER FREEZES AT 0°C = 32°F AND BOILS AT 100°C = 212°F; THESE ARE BOTH EXAMPLES OF THE GENERAL FORMULA GOING FROM °C TO °F:

$$X°C = (1.8X+32)°F$$

FOR EXAMPLE, 20°C = (36+32) = 68°F.

A RELATED (BUT SIMPLER) FORMULA MUST BE USED FOR **TEMPERATURE CHANGES**. FOR EXAMPLE, AN INCREASE (OR DECREASE) OF 5°C IS EQUAL TO AN INCREASE (OR DECREASE) OF 9°F, WHICH IS AN EXAMPLE OF THE GENERAL FORMULA FOR INCREASES OR DECREASES:

$$+X°C = +1.8X°F$$

FOR EXAMPLE, +4°C = +7.2°F.

NOTE THAT A ROUGH APPROXIMATION IS THAT A CHANGE OF X°C EQUALS A CHANGE OF 2X°F. ALSO NOTE THAT YOU CAN REVERSE THESE FORMULAS TO GO FROM °F TO °C:

$$X°F = (X-32)/(1.8)°C \quad \text{AND} \quad +X°F = +X/(1.8)°C$$

TRAGEDY OF THE COMMONS
SITUATIONS SUCH AS TRAFFIC CONGESTION, OVERFISHING, AND POLLUTION IN WHICH INDIVIDUAL SELF-INTEREST DOES NOT LEAD TO A GOOD OUTCOME FOR THE GROUP AS A WHOLE; ECONOMISTS EMPHASIZE THAT THESE SITUATIONS FEATURE COMMON PROPERTY, WHICH CAN BE MISUSED BY SELF-INTERESTED INDIVIDUALS: 135–44, 148, 160, 187, 191

U

ULTRAVIOLET (UV)
SEE **ELECTROMAGNETIC RADIATION**

UNCERTAINTY
A MAJOR CHALLENGE CONCERNING CLIMATE CHANGE IS UNCERTAINTY ABOUT HOW HIGH TEMPERATURES WILL GO AND WHAT THE RESULTING IMPACTS WILL BE FOR PEOPLE AND THE PLANET: 120–32

W

WATER CYCLE
EVAPORATION, PRECIPITATION, AND OTHER ASPECTS OF THE MOVEMENT OF WATER AROUND THE PLANET: 90, 96–97

WATER VAPOR
SEE **GREENHOUSE GAS**